THE WISCONSIN FARM THEY BUILT

THE WISCONSIN FARM THEY BUILT

· Tales of Family & Fortitude ·

COREY A. GEIGER

Foreword by Jerry Apps

THE
History
PRESS

Published by The History Press
Charleston, SC
www.historypress.com

Copyright © 2023 by Corey Geiger
All rights reserved

First published 2023

Manufactured in the United States

ISBN 9781467152747

Library of Congress Control Number: 2022950074

To grandparents who grace us with the most precious gift of all—time.
Elmer and Julia showered me with both of their gifts of time and precious stories.
As a result, a lifetime of love, laughter and lessons learned
remain with me every day.
For that, the world and I are better.

CONTENTS

FOREWORD

As a longtime rural historian, I found this Corey Geiger book fascinating, informative, entertaining, carefully researched and well written. This is a very personal book. Corey Geiger tells the story of his maternal grandparents, Elmer and Julia, from the time when they were kids until their passing.

It begins with a sad story. The date was September 14, 1932. Elmer was but fifteen years old when his forty-seven-year-old mother was hit by a train and killed, leaving behind young Elmer plus two brothers and three sisters. Elmer's little eight-year-old sister, Dorothy, cried on learning of her mother's death, "Who will take care of me?" In those days, in the midst of the Great Depression, it was common for motherless children to be put up for adoption.

"Don't worry, I'll take care of you," Dorothy's brother Elmer said. Thus we learn much about Elmer, who grows up "taking care of things," including taking over his father-in-law's Manitowoc County, Wisconsin farm when his father-in-law died.

We learn what farm life was like before electricity, tractors, grain combines, corn pickers and forage harvesters became common on farms. It was a time when farmers depended on one another for such tasks as threshing grain, filling silos and making wood, as many farmers depended on wood to heat their homes.

Grandpa Elmer, as did many dairy farmers of the day, also raised hogs in the 1940s. In those days, the local cheese factories were happy to give farmers a byproduct of cheese making—whey. Hogs thrived on whey. By the 1950s, farmers in northeastern Wisconsin were unhappy with the prices they were receiving for their pigs. So, they organized a Wisconsin Feeder Pig Marketing Cooperative. Likewise, with the help of the local county extension agent, Truman Torgerson, and Rudoph Froker, dairy marketing specialist at the University of Wisconsin–Madison, farmers organized Lake-to-Lake Dairy Cooperative as a place to sell their milk. It wasn't easy. Not everyone wanted to see the cooperative organized, but with its formation and the development of a Grade A milk market, farmers received more money for their milk.

Elmer bought his first tractor in 1949—a Model WD Allis-Chalmers. Corey Geiger now takes us on a brief tour of Allis-Chalmers (West Allis, Wisconsin), a company with roots going back to the mid-1800s. Formed as Allis-Chalmers in 1901, the company became the largest employer in Wisconsin. Among other products, Allis-Chalmers became an important manufacturer of nuclear equipment for the Manhattan Project, officially created in 1942, which led to the development of nuclear bombs. Allis-Chalmers profits peaked in 1980 and then began sliding into negatives. In 1987, the company filed for bankruptcy.

Different from many farmers who had farmed with horses and loved their horses as if they were family members, Elmer loved his orange Allis-Chalmers tractors. In those days, tractor color was important to farmers—perhaps it still is. Green was for John Deere, red for Farmall (International Harvester), orange for Allis-Chalmers and prairie gold for Minneapolis-Moline.

Elmer and his wife, Julia, lived interesting lives filled with both joy and sorrow. Elmer once said, "Farming may not be the highest compensated profession, but the children we raise and the integrity we instill in the next generation is worth more than any paycheck." Elmer died on July 14, 2008. His wife, Julia, died on September 27, 2011.

Corey Geiger does a masterful job not only capturing the rich farm life of his grandparents, but he also captured the essence of a way of life—living on a family farm—that is fast disappearing.

JERRY APPS is author of *Wisconsin Agriculture: A History* and several other fiction and nonfiction books about agriculture and rural community history. He has also done six hour-long TV shows for PBS Wisconsin about early farm life and country living.

ABOUT THE COVER

A dog is a man's best friend," entered the world's lexicon when King Frederick of Prussia first proclaimed the statement in 1789. While Elmer Wilfred Pritzl certainly liked his farm hunting dogs, a more fitting statement for the city boy turned farmer would have been, "A tractor is a farmer's best friend." That's because Elmer, the central character in this book, started his farming career in an era when horsepower literally came from four-legged animals. The purchase of his WD Allis-Chalmers tractor in 1949 transformed his work life and made farming a far more predictable endeavor with a more consistent mechanical power source. Eventually, this tractor became known as "Grandpa's Tractor."

On an early autumn day in September 1999, famed photographers Julie Lindemann and John Shimon arrived on the six-generation family farm with the goal of picturing that WD Allis-Chalmers with Elmer's granddaughter Angela, on the fiftieth anniversary of the tractor's creation and service on the same farm—truly a unique story into itself. Always a stagehand during the first decade of these photography endeavors, I set up the tractor and wagon on a hill overlooking the family farm. That's when Julie and John went into action. After they took the first shot with my sister, the author of this book interrupted and said, "This photo would be even better if we had the original owner."

"It would," quipped back John.

"I'll give him a call." Both Julie and John were amazed at my quick response, as five decades had passed since the tractor's arrival on the homestead and they knew the original owners would have been deep into the twilight of life.

Wanting to be certain that my grandparents Elmer and Julia would make the two-mile journey to the farm, during the short phone call I simply asked if they would put on work clothes and come on out to the farm because we needed extra help. It wasn't much of a sell, as they enjoyed the trip, one they made thousands of times before in their retirement.

On their arrival, I shared the details of the project. Grandpa Elmer was hesitant, but Grandma Julia was all-in. This moment presented an opportunity to document family history on the farm that had passed through her mother, Anna, to Julia and then to Julia's daughter Rosalie. Each found a husband and a fully willing partner to run the farm. On this day, Julia charmed Elmer into being a model. The farm's matriarch and patriarch climbed into the farm pickup truck.

When we arrived on the hill overlooking the farm buildings, the expressive Julie and John instantly began talking up Elmer's Allis-Chalmers tractor. That encouraged Elmer to walk over to his tractor and tell a story and prepare for the second photograph of the day. The first photograph was with star model Angela, who had been a focus of Julie and John's photography sessions over a span of ten years.

With the fall sun setting in the west and lighting up the tractor on the hill nearly perfectly, John asked Julia if she would like to be in the photo. Without saying a word, Julia walked up to Elmer, interlocked arms with her husband of sixty-one years and looked at the camera. The shutter clicked on the eight-by-ten-inch Deardorff camera that was designed to capture historic images. Julie and John knew they had just filmed a keeper. That third photo hung in the homestead house and made an appearance only for the obituary of Elmer Wilfred Pritzl. It is now available for the world to enjoy as it graces the cover of this book.

Cover photo by Julie Lindemann and John Shimon
Shown from left: Angela (Geiger) Zwald, Julia and Elmer Pritzl

ACKNOWLEDGEMENTS

E ven for a journalist with nearly three decades of professional experience, writing a book can be a daunting proposition. Such was the case with my first book, *On a Wisconsin Family Farm*. However, a metamorphosis of sorts took place in me after that book's release in March 2021.

That evolution occurred because of thoughtful interactions with readers and store owners during the journey of making visits to over two hundred stores through Wisconsin and portions of Illinois, Iowa, Michigan and Minnesota and in making dozens of presentations to groups throughout the Badger State. In each of those snapshots in time, readers, both young and old, shared their family stories with me and often offered that my first book sparked conversations about tales long tucked away in their family's mental archives.

Those thoughtful comments both inspired and motivated me to double down and write the next book.

Portions of this book previously appeared in the Homesteader's Hope series that ran over the course of two years in the *Brillion News*, a weekly newspaper that is based in the locale of my formative years. Editor David Nordby provided valuable critique, as did the readers, offering their advice to improve and expand the stories and cheer on my storytelling.

My wife, Krista Knigge; my mother, Rosalie Geiger; my aunt and uncle Annie and Bob Krueger; my sister, Angela Zwald; my uncle Albert Geiger; and my father-in-law, Pete Knigge, carefully reviewed each chapter and offered their suggested enhancements. With the final text knitted together,

Katelyn Allen, a talented editor and co-worker on the *Hoard's Dairyman* editorial team, pored through the manuscript and offered thoughtful advice. Kelly Wood, a longtime co-worker and valued grammar specialist, tightly reviewed the captions.

C. Todd Garrett, art director for the family of W.D. Hoard & Sons Company publications, scanned all the images, removing the ink notations on the front of the pictures originally written by Grandma Julia Pritzl. However, as many a reader shared with me, "At least someone recorded the names of those people!" Of course, that comment was made because many families have historic images, but without written identification, the known names of those faces went to the grave. My gratitude also goes to John Rodrigue and his teammates at Arcadia and The History Press. Their guidance has been most beneficial, and their willingness to consider the many unique ideas on my book has been appreciated.

On October 1, 2022, just a day prior to a twelve-plus-hour-per-day work week unfolded at the Fifty-Fifth World Dairy Expo, I had the good fortune to attend the forty-seventh anniversary celebration of Dregne's Scandinavian Gifts in Westby, Wisconsin. Jana Dregne, store owner of nearly five decades, hosted a group of authors and set up tables at her store entrance. With limited space, two authors shared each table. As fate would have it—fate encouraged by Jana, I might add—she placed yours truly at a table with author Susan Apps-Bodilly. We conversed with each other and readers alike on that day. As the four-hour event came to a close, Susan kindly purchased two copies of my first book and asked that one be inscribed for her parents.

About two weeks later, I received the most thoughtful of emails from Susan's father, Jerry Apps. By the end of that email correspondence, and upon review of the manuscript for this book, the role model of role models for Wisconsin authors agreed to write a foreword to my book. For that I am deeply humbled and most grateful.

PROLOGUE

In the hit song "The House That Built Me," country music singer Miranda Lambert shares a story of returning for one final time to her childhood home. While that house is no longer owned by her family, the home once again comes to life as Lambert instantly returns to her upbringing, walking its hallways and recalling the moments in time that created the woman she has become today.

Unlike the house featured in that song, our house and farm remain in the family some six generations later. However, just like Lambert's song, our house and farm built our entire family, including yours truly.

A major verse in our family's farm song, "The Farm That Built Me," began to take shape when Elmer Wilfred Pritzl entered the scene in 1937. Not owning a vehicle of his own, Elmer borrowed a buddy's car to take the farmer's daughter, Julia Burich, on a date. A man with little means, having lost his mother five years earlier, Elmer surveyed the scene and drove past the farm, not believing that could be the homestead of Julia's parents, John and Anna Burich. Elmer then proceeded up the driveway. The next year, the couple married. Eight months later, Julia's father passed away.

Having lost his own mother, who also had the name of Anna, in 1932, Elmer knew what it meant to go through life without a parent. In the days leading up to his father-in-law's funeral, his mother-in-law, Anna Burich, asked Elmer to walk her into church. As the days and months continued to pass, Anna Burich was growing more and more impressed with her son-in-law Elmer, who was a foreman at the nearby foundry, supervising men double and even triple his age. As the summer unfolded, and with no male blood relatives left in her life, Anna asked her city boy son-in-law, "Would you run my farm?"

Elmer jumped at the opportunity and left what had become a promising career at the foundry. Mentors in his life were stunned by his switch of careers, as those mentors had hoped the boy who graduated from high school at sixteen would go on to become a doctor, lawyer or engineer. Elmer blossomed in his new vocation as a farmer. For his wife, Julia, it afforded her the opportunity to move back into the home in which she was born. By the time of Julia's sixty-second birthday, she had lived at another address for only eight months of her entire life.

And so, the fourth generation of family farm life unfolded as Elmer and Julia purchased and ran the farm. Even though Elmer and Julia would become the farm's new shepherds, Anna Burich never left the property and literally lived with the couple until her dying day on April 5, 1951. That's the day Anna suffered a stroke during breakfast at the very table where the family matriarch developed the succession plan for the family farm. The uniqueness of having three generations living under the same roof encouraged the transfer of a multitude of family stories from Anna to Elmer and Julia and then to succussive generations.

As the years ensued, Louis Pritzl Sr., Elmer's father, also came to live with the couple for two-month stints. A bachelor for the rest of his life after his wife's passing in 1932, Louis developed the plan to sell his home and live with each of his six children in short durations. That continued until his death in 1970. While on the farm, Louis contributed to construction and repair projects. While getting to know their grandfather, Elmer and Julia's younger children came to think that all retired people cashed their social security checks at the local tavern. Years later, those children learned that to be untrue. Indeed, German-blooded Louis liked his beer and his card games, and that's the very reason Elmer largely swore off both as an adult. Elmer knew the pitfalls of drinking and largely had no time for it.

Impeccable storytellers, Elmer and Julia Pritzl began to transfer their stories at the dinner table in 1981 when daughter Rosalie and her husband, Randy Geiger, became the fifth generation to run the family farm. Their grandson, the author of this book, was a sponge, soaking up the recollections as Grandpa Elmer and Grandma Julia came to the farm four, five, six and even seven times a week some summers. The stories were told not only in the house but also while making firewood, hay and lumber and doing a host of other farm chores. Moreover, Julia, was an expert secretary, documenting both the photos and finances. She was a product of both the Great Depression and the Greatest Generation, and because of it, she saved everything that might have an ounce of value.

To bring their stories to life, I use narrative nonfiction to share historical events of Wisconsin, the United States and the world that were also taking place. The book is written as a series of short stories. Elmer, Julia and a host of relatives are characters in some good tales that unfolded in real life. Those oral histories are now being told in this book, *The Wisconsin Farm They Built: Tales of Family and Fortitude.*

1

KILLED BY A TRAIN OVER MILK

Adulthood. When does it really begin?

Legally, there are benchmarks that indicate adulthood is on its way or that it's even arrived by the letter of the law.

There are the moments teenagers pine for—the driver's license at age sixteen and the high school diploma that arrives around eighteen years of age, indicating one's primary education has come to fruition. For young men, age eighteen also brings forth the legal obligation to sign up for Military Selective Service with Uncle Sam. That same birthday activates the Constitution's Twenty-Sixth Amendment granting citizens the right to cast a ballot in elections held for America's great republic. Of course, there's the long-anticipated day for many a young person who has long aspired to have a legal alcoholic drink on his or her twenty-first birthday.

Despite these moments in time, the lines remain fuzzy between youth and adulthood, as many adult children use Mom and Dad's home for base camp much later in life than prior generations. And so, many young people cannot definitively declare the day adulthood arrived, the moment when the full weight of adult responsibility began placing pressure on their shoulders and at times causing sleepless nights.

That wasn't the case for Elmer Wilfred Pritzl. Even though the aforementioned modern-day milestones had yet to arrive in everyday American life, there were no fuzzy gray lines for this transition to adulthood. The black-and-white moment of adulthood smacked the fifteen-year-old boy right in the face and left an internal scar Elmer carried the rest of his earthly life. It arrived with a great heap of pain on Wednesday, September 14, 1932.

Taken on the Eigenberger Homestead in 1932, this was the last image of Louis and Anna Pritzl together with their six children. Anna would be killed by a train in September. *Author's collection.*

The events of the early autumn day forever reshaped Elmer's life and ultimately made him work harder than any boy his age in that area from that point forth. Elmer went from paperboy and farmhand to working in a foundry in short order.

ELMER MADE THE ID

"That's my mom!" Elmer screamed out loud standing on top of a bench inside Brillion's train depot.

It was about six o'clock that evening, and the city's paperboy had been out delivering the *Green Bay Press Gazette* door to door just as he did every evening after school.

This day was different, however. The evening Chicago-Northwestern train left the depot and headed north just like it had done hundreds of times before. Moments after the train left the station, its steam whistle shrieked out the most unusual warning that mid-September day. In fact, the whistle wailed almost nonstop. Then, like a passing storm, it fell silent. After some time, the train once again began belching sound from its whistle as the locomotive did the unthinkable—it began to reverse course back to Brillion's train station.

This unusual turn of events prompted the highly scheduled Elmer to deviate from his paper route that evening. On arriving at the train station, Elmer had no choice but to climb up on the bench to get a better view over the heads and shoulders of the assembled throng, who, by that time, were attempting to identify the disfigured woman's body. No one knew who she was, even though the lady had walked their streets for decades. The body was a bloody mess.

On getting a better vantage point, everything flashed before his eyes. "The dress, that's the dress my mom wore when she kissed me on the cheek as I left for school," Elmer instantly thought to himself as he gazed on the mangled body. Then came the most terrible awareness he ever experienced. "That's my mom!"

He jumped down from the bench, ran out of the building and headed straight for home.

Elmer had solved the town's unanswered question. Everyone in the depot on that September day instantly knew the maimed body belonged to Anna Pritzl.

THE VERBOTEN QUESTION

Identifying a body is a terrible responsibility—incredibly more so if it's your parent.

Given the trauma of that day, the entire Pritzl family, and all the children and grandchildren who came thereafter, never talked about September 14, 1932. The day was pushed deep into a lockbox, presumably never to be reopened again for discussions.

Then, one day, a grandson did the unthinkable. He asked the question of questions after having read a newspaper account on the events. "Grandpa, can you tell me about the day your mom died?"

Grandma Julia looked in near horror from her rocking chair as she sat next to her husband, Elmer. "That's something we do not discuss," she said sternly and picked up the rocking pace as if propelling energy from the train's locomotive on that September 1932 day. The energy and look were so stern that the thirty-four-year-old grandson fell silent.

With a tear rolling down his cheek, the eighty-eight-year-old man turned to Julia, placed his hand on top of her hand resting on the rocking chair arm, lovingly looked her in the eyes and uttered, "It's time."

He turned back to me and slowly began to talk. The words didn't flow like normal. The pain of the day rushed back into Elmer's body like he was in

By the time Elmer Pritzl (*right*) told the story of his mother's death to any of his children or grandchildren, he was deep into his eighties. Unable to complete the narrative due to its emotional nature, his wife, Julia (*right*), finished telling the story. *Author's collection.*

the television show *Quantum Leap*, in which the main character flashes back in history and attempts to prevent calamitous events. At first, Elmer set the stage in broken words and phrases. However, it also was abundantly clear that he remembered the events from September 1932 with vivid detail.

Elmer began to tell the story.

Just as she had done hundreds of times before, Elmer's forty-seven-year-old mother set out that day down the railroad tracks of the Chicago-Northwestern train line that ran through Brillion. It was a normal trip for her from the family's home in Brillion across the street from St. Mary's Church near the intersection of Center and Custer Streets.

Fresh milk provided great nutrition for Anna Pritzl's growing children, who included (*left to right*): Mary, Louis, Veronica, Elmer and Dorothy, being held by Anna. *Author's collection.*

Getting fresh milk in those days was a big deal, and the John Mulhaney farm was a great source for Anna's six children, then ages six to eighteen, and her husband, Louis Pritzl Sr. However, the errand this day ended any way but typically.

"I had been delivering newspapers for the *Green Bay Press Gazette* after school. That's when I noticed the train backing up into the depot. It wasn't normal. Its whistle was blaring sound nonstop as it backed up. That caused me to head to the depot," continued the fourth born in a family of six children. Tears rolled down his cheeks during our 2006 interview. "I noticed a large crowd had formed," continued Elmer, now telling the story with his eyes closed, as the pain was too piercing to keep them open and witness the looks of those in the room.

"It took me some time to wade through the crowd. People were murmuring possibilities of the deceased. Since I was only fifteen and too short to see, I jumped up on a bench," he said, pausing for a long moment of reflection, all with eyes still closed.

"The person's body was lying on the depot floor. I saw Dr. H.F. Smith examining the woman," said Elmer. "'That's my mom!' I exclaimed, and the crowd turned and looked at me as I stood on the bench with my newspaper bag in tow and a baseball cap on my head. I knew it was my mom because I recognized her clothes," Elmer said, now visibly crying, as if September 14, 1932, was happening all over again.

Then he said, "I ran home and cried."

With that statement, I asked no more. Grandpa Elmer was clearly done talking. That's when his wife, my grandmother Julia, continued the conversation.

"The newspaper recount didn't include some of the details," she said. "The train engineer tried to stop the train, as he could tell her boot was caught in the track," Julia added. "In those days, women wore lace-up boots with many eyelets. Her boot either got caught in the track or railroad spike and she couldn't unlace it fast enough."

The Newspaper's Account

Dateline: September 16, 1932, the *Brillion News.*
Headline: Mrs. Louis Pritzl Meets Tragic Death

"At about 6 o'clock on Wednesday evening, while returning from the John Mulhaney farm with milk, and as she was crossing the Chicago & N.W. railway tracks a half mile west of here [Zander Press], Mrs. Louis Pritzl was struck by No. 153 northbound passenger train, due at 5:45.

"The engineer attempted to stop his train when the woman appeared not to have heard the warning whistles and approaching danger, but it was impossible.

"When the train backed and the crew rushed to the scene, the woman was found to be horribly mutilated about the head and apparently dead. The remains were brought back to the depot and Dr. H.F. Smith summoned. He made a hurried examination but announced that no further medical aid would be of avail, stating that death had probably been almost instantaneous.

"Father Kraus, pastor of St. Mary's congregation, and the children at home were called. Mr. Pritzl, when the accident occurred, was at work at the Charles Giese Farm and did not learn of it until sometime later.

"After a half hour, during which train officials made a report of the tragic happening, the train proceeded.

"Friends and neighbors quickly gathered at the depot and assisted Father Krause in consoling and caring for the grief-stricken children. The sorrow

Annie and Simon Eigenberger (*seated*) are shown with children Margaret "Maggie," William "Bill" and Anna. When Anna died in 1932, her oldest son, Arthur, was working on the Eigenberger family farm run by her siblings Bill and Maggie—neither of whom ever married. *Author's collection.*

spread over the entire community, and sympathy for the bereaved family was expressed by all.

"District Attorney Edward S. Eick was also present at the station and called Sheriff Gerhard Jensen. The latter conducted no inquest, however, as it was clearly evident that it had been an unfortunate accident.

"The body was then taken in charge by undertaker C.F. Koch, and transferred to his undertaking parlors, and later to the home on Center Street.

"The decedent, who, before marriage was Miss Anna Eigenberger of Grimms, was 47 years of age. She was united in marriage to Louis Pritzl in 1912 at Clark's Mills. They have resided in this city practically since that time.

"Surviving the unfortunate woman are the husband and six children— Arthur, Mary, Veronica, Elmer, Louis, and Dorothy, all at home except Arthur who is employed at the Wm. Eigenberger farm near Grimms. Mr. Eigenberger is a brother of the deceased and at his home also lives a sister, Margaret.

"The funeral will take place tomorrow [Saturday] at 9 a.m. Services to be conducted at St. Mary's Church. Rev. Krause officiating. Interment will be in the Catholic Cemetery."

It may be hard for some to believe a pedestrian could be killed by a train. To this very day, railroads still deserve extra caution when crossing tracks. According to the National Highway Traffic Safety Administration and the Federal Railroad Administration, a motorist is twenty times more likely to die in a crash involving a train than a collision involving another motor vehicle. That's based on modern-day statistics, as little data from the 1930s exists, a time when the ratio of automobiles to trains was much lower. Also, this data does not consider pedestrian traffic.

I'll Take Care of You

What would become of the Pritzl children?

While America was mired in the depths of the Great Depression, these Pritzl kids had a worse lot in life—no mother. Days after her mom's death, Dorothy, the youngest of the six children, cried her heart out after hearing adults talking about what to do with the six Pritzl children. In those days, it was unusual for a single father to care for a brood of youngsters without a woman.

"Who is going to take care of me?" the eight-year-old Dorothy exclaimed out loud.

Dorothy feared she would be put up for adoption. Knowing this, Elmer picked up his eight-year-old sister, sat in a chair and set Dorothy on his lap in the fall of 1932.

"Don't worry, Dottie, I'll take care of you," the teenage Elmer assured his little sister wiping away her tears. In calling her Dottie, not Dorothy, the older brother gave an added measure of assurance.

While his father, Louie, held firm against all the assembled do-gooders who thought they could best raise his family by parsing the task out to the others via adoption, Elmer knew he had to become an adult at that very moment. That's because Elmer also knew that a significant amount of their father's paychecks had been routinely transferred to local tavernkeepers over the years, as their father liked to drink his share of good German beer and play every card game under the sun. That caused the elder Pritzl to shirk some of his daily responsibilities.

In the months after his mother's death, Elmer learned just how troubling the situation had become and why his paper route money never balanced at month's end despite keeping impeccable ledgers. His mother had been pulling from that newspaper stash to buy food for the family. As Elmer entered adulthood, it was his father's drinking problem that caused him to largely swear off alcohol consumption his whole life.

Back to Elmer's promise to his little sister Dorothy. Could he live up to it?

Elmer had been hitting the books hard and had already skipped two grades by testing out of those class levels with the dire economic circumstances of the 1930s. Now, with a deceased mother, the financially pinched family trekking through the Great Depression was in desperate need of money, and Elmer aimed to land a full-time job as quickly as possible.

Elmer graduated with Brillion High School's class of 1934 at age sixteen. Then he took a job at the Brillion Iron Works, climbing to the rank of foreman before turning twenty.

Elmer squirreled his money away, accounting for every penny in a ledger.

Decades after Anna's death, Dorothy told her siblings of Elmer's good deeds and ensured her caring brother received the first payout from the family estate. Shown here are, seated (*left to right*): Veronica Voss (née Pritzl), Dorothy Kubale (née Pritzl) and Mary Lashua (née Pritzl); standing, Elmer, Louis Sr., Arthur and Louis Jr. *Author's collection.*

True to his word, when Dottie needed an emergency appendectomy in March 1937, Elmer covered the surgery. When Dottie aspired to become a registered nurse, Elmer paid a major share of his little sister's tuition because he wanted for Dorothy what he didn't have—a college education. Elmer's dreams to become an engineer or doctor blew away with the dust and dirt of the Dirty 1930s and the Great Depression.

He would not have that for his little sister. He took care of Dorothy and largely told no one except his wife, Julia, who loved Elmer even more for his kind deeds.

Years later, after the Pritzl family home in the center of Brillion was sold, Dorothy quietly told her siblings of Elmer's deeds. On hearing the actions of their rather private brother, the five siblings matched Elmer's style and made sure he was the first to receive money from the estate.

The gesture touched Elmer's soul.

Elmer, in turn, took the money and bought a Legacy Clock crafted by Seth Thomas that chimed a beautiful tune at the top of each hour and at each quarter hour. That song served as Elmer's reminder of the treasured relationship with his sister Dottie, who graduated as a registered nurse and eventually became a supervisor at Green Bay's Bellin Memorial Hospital.

The Legacy Clock still chimes in the home of Elmer's youngest daughter, Annette Krueger.

While the catastrophic death of Anna Pritzl accelerated Elmer straight into adulthood, he did his best to ensure his younger siblings had some resemblance of a childhood and that they could live out their dreams as adults.

2

THIS DOG BECAME A TREE

Bong…Bong…Bong," tolled the brass church bell.

"Hoooowl…Hoooowl…Hoooowl," echoed back the dog.

Without fail, whenever that church bell rang, the dog tied directly across the street would yowl something fierce. It was clear that the sound bothered the dog—really badly.

The bell was located high in the steeple of St. Mary's Church situated at the corner of Custer and Center Streets in Brillion. To the area's Catholics, its clanging sound meant that Sunday service would soon start.

Between the bell's repetitious tolling each week and the dog's counter howling, Brillion had a noise war on its hands.

Decision Time

"You have to do something about that dog, Elmer," exclaimed a towering figure of a man wearing a black shirt, black pants and a white collar. "He cannot howl like that when Mass is taking place!"

It was a command delivered by a highly respected citizen living within Brillion's city limits—Reverend August Garthaus. Reverend Garthaus served as St. Mary's Parish priest for twenty-two years. The church's centennial directory some fifty years later included the words "authoritarian nature and zeal" as descriptors for Reverend Garthaus. One could only imagine how twelve-year-old Elmer Pritzl received the stern message that Saturday morning as the parish priest towered on the Pritzl front porch.

When Reverend August Garthaus (*standing right*) issued the command "Do something about that dog," Elmer (*boy seated far right*), who was in the eighth grade, jumped into action. *Author's collection.*

Not only was Elmer an altar boy for Reverend Garthaus, but also Elmer's parents, Anna and Louis Pritzl, were devout Catholics. To be candid, Elmer was mortified. Although still a preteen boy, Elmer knew Father Garthaus was issuing an order, not a request, that the young lad had better follow up on—right quickly, too.

Garthaus was a unique character in his own right. He was a rather rotund fellow who fancied porterhouse steaks from the local butcher shop. Elmer's mother volunteered her son for Reverend Garthaus's food delivery, as they lived just across the street. While he wasn't a fan of pets, Reverend Garthaus did own a parrot named Polly. Polly the parrot spoke a few second-rate words within the confines of the rectory, as Elmer knew firsthand, having frequently made food deliveries to the priest's residence. However, that talking parrot was not disturbing the peace like Elmer's dog. So, Elmer decided to remain mum about the pet parrot.

"With me about to be a teenager later that summer, it would have been the beginning of the Great Depression," recalled Elmer of the dubious occasion. "When the church bell would ring before Benediction, my dog

The puppy shown on the left would grow up to become Brillion's howling hound. Elmer, shown in front of the family home, would later trade that dog for a tree. *Author's collection.*

would howl," said Elmer, making no qualms about this issue. "And boy could he howl," he said of his dog.

What could Elmer do with his dog?

Elmer approached Joe Braun, who owned a farm just east of Brillion on Highway 10. Elmer worked there whenever he was not in school or delivering papers for the *Green Bay Press Gazette*.

"I'll trade you a good rabbit dog if you trade me a hickory tree about the thickness of my pinky finger from your woods," pleaded the much younger Elmer as he showed his finger to the elder Joe Braun. Braun, knowing the boy's plight as a fellow Catholic, made the trade.

"I didn't know what I was doing, but I knew I had to get rid of my dog and I wanted to plant a hickory tree at my house. I liked planting trees and working with wood," Elmer reasoned.

"That summer was awfully hot, so I watered that little hickory tree every day. Despite the extreme heat and dry conditions, that little tree made it."

"On the flip side, Joe Braun got one fine hunting dog," said Elmer, who noted the dog was far away from the church bell.

"Part of the area where Brillion's T&C Grocery Market stood for many years was our family's lawn back in those days," said Elmer of the location of his hickory tree. "Years later, city leaders demolished the entire block in 1977. The city paid construction crews to move the houses away to the country or else they would be smashed," he said. "Our family's house got destroyed. Other houses went east a distance. Some of those houses eventually were moved a second time to Forest Junction," Elmer went on to say. But the tree Elmer planted in 1929 remained in its place.

While Elmer's hickory tree survived the first bulldozers in 1977, four decades later, Brillion's city officials levelled the retail hub to return it to a residential development. The second time around, Elmer's tree was taken out for the sake of progress.

IT WORKED FOR PAVLOV

Years later, after Elmer married Julia Burich, he began running the Burich Family Farm nestled midway between Brillion and Reedsville. To Elmer's delight, the Burich farm had a great deal of wooded land. The delight part came into the picture because Elmer loved to hunt.

Elmer wanted a partner to assist him in his hunting expeditions. He was short on funds though, so the dog had to be cheap.

Farm dogs and hunting dogs are often one and the same. Elmer is shown here in the late 1970s with his last farm hunting dog, Buddy. *Author's collection.*

His neighbor Steve Foreyt had a beautiful Irish setter. There was only one problem with that dog—when the canine saw a gun, much less heard it fire, it ran and cowered under the porch. Elmer learned of the situation after church one day while talking with the older farmer.

When Elmer offered a piglet for the dog, Foreyt quickly made the trade. Foreyt, who farmed on nearby Manitowoc Road, thought he made out like a bandit getting a piglet for a hunting dog that wouldn't hunt. And Foreyt easily would have won that trade if it hadn't been for Elmer's grandiose plan.

What would Elmer do?

An avid reader, Elmer had learned about Ivan Pavlov's Nobel Prize–winning work from 1904. In that work, Pavlov noticed that dogs would salivate whenever he entered the room. The reason? The dogs had developed a conditioned response and associated Pavlov's entrance with feeding time.

Through a series of experiments, Pavlov later got those same dogs to salivate whenever he rang a bell. That's because after Pavlov rang the bell, he fed them. For the dogs, the bell and its ring became associated with dinnertime.

How did Elmer make use of Pavlov's discoveries?

He fed that Irish setter twice a day. To keep close watch on the dog, he tied it on a chain halfway between the family's home and his dairy barn.

Instead of Pavlov's bell, Elmer used his shotgun. He fired off a shot, then gave the Irish setter its food.

In about two weeks, that dog was jumping up and down in anxious anticipation of food whenever he saw Elmer with his shotgun.

"He knew it was dinner time," smiled Elmer. "And in a few short weeks, I knew that I had a hunting dog."

Elmer and that Irish setter would become devoted hunting partners for well over a decade, as the dog was bred for bird and rabbit hunting.

And this time, Elmer's canine companion wasn't disturbing the peace and tranquility of Brillion's residential neighborhood. Instead, the nearby parish priest, William Koutnik of St. Mary's of Reedsville, had become a fan of Elmer's dog.

So much so that he even asked, "Elmer, when can we go rabbit hunting?"

3

THE SIXTEEN-YEAR-OLD GRADUATE

"Elmer needs a challenge," said the Mother Superior Franciscan Sister who had called a meeting with Elmer's parents, Louis and Anna Pritzl, in the spring of 1927. "He's so bright."

On the day of the meeting between his parents and schoolteachers, Elmer would have traditionally been promoted from the fourth to fifth grade.

"We recommend that he proceed immediately to the seventh grade," said the leader of the team of assembled nuns. The group who instructed students at St. Mary's Catholic Grade School in Brillion had met earlier and believed Elmer was a budding child prodigy who could skip two grades of instruction.

Elmer's mother, Anna, was beyond flattered by the Catholic nun's assessment of her son at the meeting held in the pristine, five-year-old school building. She knew her son to be clever and was ready to fully support the idea. After all, she had been the only one endorsing the report cards with her signature after Elmer brought them home. However, his father, Louis, had some serious reservations. He expressed those matters immediately, in a rather blunt fashion.

"He's such a young boy," spouted Louis, who knew his August-born son was already one of the youngest in his class even without this super promotion.

"If you move him up two grade levels, he will have no friends and fellow schoolchildren will resent him," remarked the gruff German, who had been a lumberjack in Wisconsin's Northwoods and could hold his own drinking

Left: Early in life, Elmer Pritzl's parents, teachers and community leaders began to realize he was a clever child. *Author's collection.*

Below: During his teenage years, Elmer worked at the Braun brothers' farm just east of Brillion. Years earlier, Elmer's father, Louis Pritzl Sr., was the stonemason who built the silo. *Author's collection.*

beer and playing cards with anyone in the area. "Plus, he will not be able to defend himself," added the mountain of a man, explaining that the other schoolchildren would be far more physically adept in comparison to the much younger Elmer. Clearly, the senior Pritzl had previously learned from the school of hard knocks about defending himself in lumber camps powered by rough and tough immigrant labor.

"How about one grade level?" retorted the father of pure German descent. It meant Elmer would skip the fifth grade. All agreed. And so, it came to be that Elmer entered the sixth grade as a ten-year-old in the fall of 1927.

HE DIDN'T WANT THE ATTENTION

After entering the sixth grade, Elmer went about his schoolwork in his same methodical and diligent manner. He went out of his way not to draw attention to himself, as he simply wanted to fit in. Trying to earn a few extra dollars, he also began delivering newspapers for the *Green Bay Press Gazette* and worked for the Braun brothers on their farm just east of Brillion on U.S. Highway 10.

Elmer's plan to keep to himself on academic matters worked until a few years later when Brillion High School principal C.H. Wileman conceived a plan in the spring of 1931. We know this story because of a rare journal entry by Elmer. While a gifted student, English did not top his list of favorite classes. That being the case, Elmer's childhood journal consisted of exactly eight pages. At the time, Elmer was a freshman, but based on age he should have been either a seventh or eighth grader.

"When I was in high school, I took a test when I was in my freshman year," wrote Elmer. "It was an Iowa Achievement Test for the seniors. Since they didn't have enough seniors, Principal Wileman asked that I take the test. As it turned out, only four seniors got a better score than me," wrote Elmer.

"Then they paraded me before the 'assembly of all the classes' and Principal Wileman told the students that he was surprised that a freshman could do so well," wrote Elmer, as the Brillion principal used a combination of tough love and embarrassment to motivate older students who could have been doing better in their studies. "And of course, I was embarrassed and wanted to run out of the gym," he concluded his short journal entry of the occasion.

In having been promoted two grades in one year, Elmer, shown here at his eighth grade
graduation, carried a rather small frame into his freshman year of high school. *Author's collection.*

NO MORE SIGNATURES

In those days, parents signed and dated report cards. Elmer's grades were rolling in strong despite his youth, and his mother, Anna, signed every one of them, *Mrs. Louis Pritzl*, in her impeccable cursive handwriting she perfected as a student at nearby St. Mary's of Clarks Mills. That included a report card from the spring semester in 1931 when Principal Wileman was boasting to the school assembly about his sharp young student.

After 1932, no signatures were ever placed on Elmer's report card again. As Elmer was entering his junior year in high school that fall, Anna was killed by a train. With the family in disarray, no one ever signed or likely even looked at another one of Elmer's report cards.

People in the Brillion community immediately stepped forward and volunteered to literally adopt Elmer as their son after his mother's death. Those Good Samaritans, who included a doctor and a lawyer, knew Elmer's father, Louis, would have his hands full with a brood of six children. With some extra nurturing, Elmer could develop into one of Brillion's shining stars.

However, Louis would have nothing to do with it. He let everyone know it, too.

Louis trudged on, taking care of six children and often shunning help out of pure pride. Despite holding down two jobs and losing his mother, Elmer's grades held steady. He ranked top among all students in a variety of classes, including biology, chemistry, economics, home mechanics, manual arts, mechanical drawing, physics, American history and world history. The report cards in those days included the student's rank among his or her peers in each class. Clearly there was no such thing as grade inflation in those days, as Elmer ranked first out of nine students in the mechanical drawing class. However, that top spot earned only a grade of "B" from an old-school instructor who employed one tough grading system.

Elmer's Brillion diploma was printed in 1934 by the Welch Company of Chicago, Illinois. It was signed by Principal C.H. Wileman; Dr. H.F. Smith, president of the Board of Education; Otto Zander, secretary; and Henry Lepple, treasurer.

In addition to Elmer's diploma, his commencement invitation for the Brillion High School class of 1934 and the senior class play bill from *The Promoters* survive to this day. In that play, originally written by Warren Beck, Elmer played the role of Vernon Clarkson, who is a student. Elmer had that role down pat.

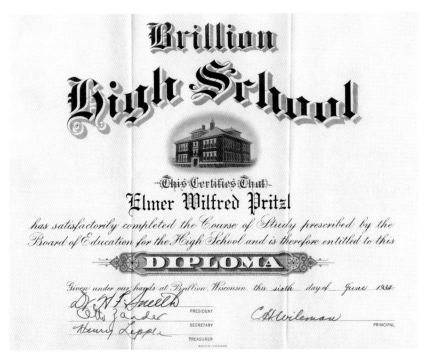

Despite being the top-ranked student in manual arts, Elmer earned a grade of B in an era without grade inflation. This report card also illustrates that no parental review of Elmer's grades occurred after his mother was killed by a train. *Author's collection.*

Several influential community leaders signed Elmer's report card, including Dr. H.F. Smith, who examined Elmer's mother the day she was killed by the train, and Principal C.H. Wileman. Both men challenged Elmer to go to college. *Author's collection.*

Even though Elmer rarely boasted of his academic achievements, he was proud to have graduated high school, so much so that he purchased a senior class ring. However, with the Great Depression deeply gripping America by 1932, he could muster up funds only for a less costly, smaller lady's ring—and that is what he wore to mark his graduation from high school.

JUST A FARMER

Clearly, Brillion community leaders were talking about Elmer. He graduated high school at age sixteen. He held down a job. He had no mother. He beat the odds that were stacked against him during America's deep economic recession.

One must imagine that Principal Wileman had talked to R.D. Peters, president of the Brillion Iron Works, on an occasion or two. That's because Peters, known as Butch by many in the area, quickly offered Elmer a position after graduation even though the plant's sales were down due to the Depression. Elmer got offered that job partially because of ingenuity and partially because his father, Louis, worked there. In those economic times, sons of foundry workers had first pick of jobs there.

Within two short years, Elmer rose to the rank of foreman at the tender age of eighteen and was supervising people twice and even triple his age. He held that role for the next three years and navigated the supervisory role well, having learned to work with others older than him while he was the youngest in his class by several years for most of his formal education.

Life for Elmer began to evolve rather quickly when he met Julia Burich in the summer of 1937. Like his sister Dottie, Julia was the youngest in her family. To take Julia, who graduated nearby Reedsville High School one year earlier, on their first date, Elmer had to borrow a buddy's car, as he was still living in the Brillion city limits and was walking to work every day.

Elmer and Julia quickly fell in love and were married the very next summer, four years after he began working at the Brillion Iron Works. After their wedding, the newlyweds rented an apartment in Brillion across from Rudolph "Rudy" Seljan's original food stand—what eventually became the famous Rudy's Café.

One year later, Julia's father, John Burich, died, and there was no one to help his widow, Anna Burich, run the 202-acre farm. Julia's sister Mary and her husband, Herb, had left that same year, as they couldn't swing it financially. That same spring, Julia's sister Agnes and her husband, George, gave it a try. That didn't fare much better.

BRILLION PULVERIZER CO. EMPLOYEES PICNIC
LONG LAKE ~ AUG 22ᴺᴰ 1936

By the time the Brillion Pulverizer Company Employees' Picnic took place in 1936, Elmer (*top row, dark shirt, arms folded*) was a foreman and pitched for the company's baseball team. *Author's collection.*

As this situation unfolded, Anna wrote a two-page letter asking her son-in-law Elmer and daughter Julia if they might consider running the farm on a rental basis with an option to buy. Having worked for the Braun brothers for many years, Elmer learned a great deal about farming and frugality. He leaped at the chance and left his job as foreman at the Brillion Iron Works.

Principal C.H. Wileman was in disbelief.

He had hoped Elmer would enroll in college to become a medical doctor, lawyer or engineer after earning enough money. Given Elmer's budding young career at the Brillion Iron Works, Wileman had come to believe that the young man would choose to become an engineer once he earned money to enroll in college.

Instead, Elmer was about to spend those hard-earned funds on a down payment for a farm and to help his little sister, Dorothy, achieve her educational goals—unbeknownst to anyone in the Brillion area at that time.

In an attempt to persuade Elmer to enroll in college, Wileman drove out to Elmer's newly acquired farm located between Brillion and Reedsville.

Principal C.H. Wileman always had high hopes for his star student, Elmer Pritzl (*standing, second row, fifth student*). However, on visiting Elmer on his farm in the spring of 1940, Principal Wileman gave up hope that Elmer would pursue a higher education. *Author's collection.*

Despite the long conversation between the two men, who clearly admired each other, Elmer showed no interest in leaving his farm. He had a wife, Julia, and a lady who treated him like her own son in Anna Burich—as Julia's mother was also living with them in the family home.

"I am more at peace here than even when I attend church," Elmer told Wileman of his chosen vocation of farmer.

"Elmer, I had such high hopes for you. I can't believe you are willing to settle with just being a farmer."

They parted ways that spring day in 1940, never to bring up the subject again.

As for Elmer's choice of being a farmer, I'd like to think that one of Wisconsin governor W.D. Hoard's most famous quotes in the late 1890s summed up Elmer's belief in his newly chosen profession: "A college cannot give a man an education and a farm cannot prevent one either. The farmer of the future must know more of certain things involving chemistry, bacteriology and mechanics, than did the farmer of the past."

And Elmer prospered, applying all his intellect to his farm. He paid off the farm's debt in five short years despite paying his mother-in-law top dollar for it.

Governor Hoard was right.

It takes brains over brawn to run a successful farm. Elmer had the brains.

4

A CONTRACT FOR LIFE

Married just fourteen months, Elmer and Julia were still newlyweds. However, the young couple was already laser-focused on their life vocation together. That was to farm on Julia's family homestead settled in 1867 by her great-grandfather and great-grandmother Thomas and Mary Burich.

That dream was about to become a reality for two reasons: Elmer had squirreled away nearly every penny possible, amassing $2,100 during the heart of the Great Depression, when many people went without a job. And even though the couple was mourning the passing of Julia's father, John Burich, the situation created a window of opportunity to take over the family business.

After receiving a letter from Julia's recently widowed mother, Anna Burich, on Friday, August 18, 1939, the couple met with her at the kitchen table two days later, on Sunday. That letter contained an invitation to run the farm. Remarkably, it took only the six ensuing months for Elmer and Julia to complete the purchase agreement on the 202-acre property.

At the onset, Anna wanted to sell just a portion of the Burich homestead to Elmer and Julia. In that first draft of a plan outlined in the August 18 letter, the couple would have to build their own home. However, discussions between the young couple and the sixty-two-year-old first-generation American-born farm woman quickly took a different direction at the Sunday afternoon meeting around the kitchen table as everyone sipped coffee.

"How about we move into the farmhouse, and you continue to live here…with us?" suggested Elmer in a polite but rather bold fashion to the lady forty years his elder. Keep in mind Elmer had just celebrated his twenty-second birthday the previous week and had married her daughter, Julia, one year earlier.

Elmer's idea had merit. Anna was living in a gigantic, five-bedroom home that also included an upstairs bathroom. That didn't even count the five large rooms on the main floor.

By this time, Anna was living there all by herself.

"Let's give it a try," she said, getting the full impression that Elmer and her daughter Julia had talked over Elmer's proposal prior to coming to the farm that day.

Elmer left his five-year job with the Brillion Iron Works. He and his bride returned to her family's farmhouse, where Julia was born just twenty years earlier on October 30, 1918. The house was constructed in 1916, making Julia the only one of her five sisters to be born in that house instead of the pioneer home. Hospital births were a luxury, and the clinic was "too far away" by the standards of the day.

A TRIAL PERIOD

Would the living arrangement and farming agreement work?

Anna may have had her doubts. She had been down this road before when her oldest daughter, Mary, and her husband, Herb Kalies, looked to buy the property. After six months, while Anna's husband, John, was still alive, Herb returned to his family's homestead near Askeaton. Herb and Mary would buy a cheese factory in southern Brown County a short time later.

After that, Anna's second-oldest daughter, Agnes, and her husband, George Kubsch, rented the Burich family farm. By the summer of 1939, it was clear that proposition wasn't going well, as there was little hay and grain stored in reserve for the coming winter.

That's when Anna reached out to Elmer and Julia with a double-sided letter. In Elmer, she saw a hardworking young man who led people twice his age while foreman at the Brillion Iron Works pouring molten iron. He had brawn and brains. They were similar traits she had seen in her husband, John.

So, Anna took the chance on Elmer. Even though he was a city boy, having grown up in nearby Brillion, he also had been a newspaper delivery boy for

On June 16, 1938, Elmer and Julia Pritzl were married, and one year later Anna Burich would invite the couple to run the family farm. Pictured here are (*left to right*): Charles Novak, Mary Lashua (née Pritzl), Louis Pritzl, Veronica Voss (née Pritzl), Elmer and Julia, Beatrice and Quiren Sleger and the little couple Ken Kalies and Janice Kubsch. *Author's collection.*

the *Green Bay Press Gazette* and held a second job working on the Braun family farm just east of Brillion. As for smarts, just five years earlier, the young man had graduated second in his class at nearby Brillion High School.

A CHRISTMAS CONTRACT

After a ninety-day trial, Anna's thoughts were confirmed by Elmer and Julia's hard work and ingenuity.

"I'll sell you the farm," said Anna one night at the dinner table over a meal she had cooked for the trio.

"However, I have terms you must agree to in order to buy it," said the full-blooded Bohemian woman who worked as hard as any man in the neighborhood.

She slid a handwritten note across the table toward Elmer and Julia. She wanted $6,100 for the 202-acre farm. That was $30 per acre, and it included 40 acres in what would years later become part of the Collins Marsh wetlands restoration project. That was a fair price at the time, especially as the Great Depression labored on and placed many a farm family in economic destitution.

Elmer had $2,100 of the purchase price saved from his job at the Brillion Iron Works. The remaining $4,000 was acquired via a loan from the Calumet County Bank.

But Anna's purchase agreement came with a few additional stipulations. Anna wanted what she had given her father and father-in-law: a home and perpetual care for the remainder of her earthly life.

LODGING FOR LIFE

Entry No. 99 on the Abstract for Title on the soon-to-be fourth generation family farm chronicles every detail of that agreement, including the aforementioned purchase price. Anna left nothing to chance. Entry No. 99 read as follows:

Elmer Pritzl & Julia, his wife, to Anna Burich agrees:

Parties of the first to provide Anna Burich board, lodging, washing, ironing, and mending, in the house on said premises, and care and nursing, in the event of illness. In the event hospital care is needed, to pay all doctor bills including those for operations, also all nursing bills and upon her death to provide her with a good Catholic burial and a monument or headstone at her grave.

Parties of the first to pay party of the second [Anna] $5 a month, if said party of the second shall eat with obligators [Elmer and Julia] at the same table. If Anna Burich shall decide that she wants to cook her own meals and live separately, she shall have the right to select two rooms, either downstairs or upstairs in the same house. Parties of the first shall put such rooms in condition so that she may be able to do her own cooking and her own housework in said two rooms, to provide same with necessary stoves. During such time while she is so living alone, she shall receive the sum of $10 a month, also all the necessary milk, cheese, butter, meat, eggs, lard, potatoes, and bread.

Parties of the first will take party of the second to and from church on all Sundays and Holy Days of Obligation, also other places as she may want to visit.

If Anna Burich should live by herself, parties of the first will furnish her with sufficient fuel and lights for her in said rooms.

Should Anna Burich decide not to live in the house now on said premises, then she shall have the right to make her home at any other place she sees fit. In this case, parties of the first shall pay her the sum of $15 a month as long as she shall live.

HOW IT WORKED IN PRACTICE

Grandchildren Jacque Pritzl and Elmer John Pritzl had come to know Grandma Anna Burich as their favorite babysitter. *Author's collection.*

That contract remained in effect for the remainder of Anna Burich's life.

Anna never exercised the $10 or $15 clause in the contract. That's because she ate breakfast, lunch and dinner with Elmer, Julia and their growing family for the next eleven years.

For Anna, the contract clearly was a fallback plan designed to protect her in the event the relationship between Elmer and Julia went sour. In many cases, Anna didn't even live by the terms she wrote for Entry No. 99, as she cooked the grand majority of the family's meals while Elmer and Julia did farm chores. In return, Elmer dutifully brought firewood, at perfect lengths and split just right, for Anna's prized wood-fired cooking stove.

And during World War II, Anna would work side by side with Julia and Elmer. Well into her sixties, Anna milked cows and cooked for threshing crews. On July 6, 1942, however, Anna woke up and declared, "Today I milk my last cow!" That happened to be her sixty-fifth birthday.

Having retired from cow chores, Anna continued her role as cook and a newfound role as babysitter. Indeed, Jacqueline "Jacque," Elmer John "Butch" and Rosalie, Elmer and Julia's first three children, got to know

Grandma Anna very well over the ensuing decade, as they all lived in the same house. Jacque arrived in May 1940, and she knew Grandma Anna as her favorite babysitter. The contract for life saw the family farm transition to the fourth generation.

5

YOU NEED YOUR NEIGHBORS

Americans have long taken pride in being a self-sufficient people. Phrases such as "I can stand on my own two feet," "I can make it on my own" or "I did it my way" have been imbedded into our lexicon. Many have come to believe in that American mystique. However, it's not always true in practice.

The truth is that early pioneers, and farmers to this very day, needed their neighbors. They relied on one another to get work done. They relied on one another to help each other get out of a bind. They relied on one another to ward off danger.

Relying on others can help get a great deal of work done—it also can bring disappointment when one falls short.

Those scenarios played out time and time again across American's countryside.

Ate Like Horses

One of the largest labor activities that required farmers to lean on their neighbors was threshing. Separating the seed or grain from the plant, a job that once took a dozen men a full day's work, can now be completed in hours by one person and a modern-day combine.

On the Burich farm, which was later owned by the Pritzls, threshing would involve crops such as barley, oats and a wide variety of clovers. The work crews were huge, as were their appetites—those workmen ate like horses.

Threshing grain and clover was a labor-intensive endeavor. *Author's collection.*

The farm family requesting the services of a threshing crew needed to feed their helpers so they kept working in their fields all day long. And since everyone's grain crop generally ripened at the same time, sometimes those who prepared the best meals received the timeliest harvest assistance.

What were the typical costs for a threshing crew?

In August 1943, Emil Kinast arrived at Pritzl's Pine Haven Farm to thresh the family's grain crop. Kinast got paid $210 to thresh the barley and oats that year.

That was just the start. Those thirteen men who came with the thresher collectively received $26 for their labor. Most of those men were neighbors of the Pritzl family, as was Kinast.

On that August day, those men were also fed two meals prepared by Julia and her mother, Anna. The kitchen would have been buzzing like a beehive the days prior to the late summer threshing event.

Julia recorded in her ledger that she purchased one hundred pounds of potatoes for $5.75. Julia and Anna peeled all those potatoes in preparation for the threshing crew's arrival. Another $4.00 was paid to nearby Prochnow's Market to supplement the family's home-grown meat, which likely included chicken, geese or pork. Garden vegetables also complemented the meal. Based on the family tradition of meals served today around the Pritzl descendants' table, creamy green beans and coleslaw likely rounded out the main course. The green beans and dill were a favorite recipe, as was cabbage-based coleslaw.

The Way to a Man's Heart

Anna knew the way to a man's heart was via his stomach. For every threshing event, Anna made her locally famous pies. In August, those pies featured black raspberries handpicked from her forty-acre family homestead.

In the days leading up to threshing, Anna also looked over her firewood inventory and made sure that her son-in-law Elmer had chopped ample firewood to ensure consistently baked pies came from her wood-fired kitchen stove. If the firewood pieces were too large, the fire would burn at a cooler temperature and bake the pies unevenly. Pieces too small, on the other hand, would burn hot and scorch the crust. Like Goldilocks, Anna was looking for firewood that was "just right." There was no doubt that Anna was persnickety on every step of pie making, as it was one of the reasons those threshing teams had made her family's farm one of its first stops.

The final purchase—beer.

Kingsbury was the beer of choice, and it was brewed in nearby Manitowoc. Julia purchased three cases for the August threshing, and those thirteen men drank every bottle. The average man with German and Bohemian ancestry would have drunk five to six bottles to quench his thirst and rehydrate.

The meals were served around Julia and Elmer's farmhouse's kitchen and dining room tables. When the steam whistle on the threshing machine blew, it signaled that dinner was about to be served.

Later that year, on September 13, August Bratz came to the farm with his threshing machine. This unit was different than the one owned by Kinast.

This time, alsike clover was the crop. Julia and Anna started cooking all over again. The menu was the same, except for desert. Apples Anna sourced from the farm's orchard filled her pies during the clover harvest.

On that day, the crew threshed 1,780 pounds of clover seed, which Elmer later sold to nearby Reinemann's during the heart of World War II. Reinemann's then resold it to the U.S. War Department, and it was crushed for oil to lubricate high-end military guns and cannons on naval ships.

Husk Corn at Pritzl Farm

The Nebraska Cornhuskers may simply seem like a well-known college football team, but corn husking was a real farm activity. Julia Pritzl penned the following writeup that was published in the society column of a local newspaper.

Mr. and Mrs. Elmer Pritzl were hosts of a corn husking bee at their farm west of the village on Monday evening. The following people attended: Mr. and Mrs. Earl Voss of Brillion, Mr. and Mrs. Herbert Kalies and sons of Cato, and Louis Pritzl of Brillion. Mr. and Mrs. Quiren Sleger and son of Menchalville, Mr. and Mrs. George Kubsch and children of Kellnersville, Mr. and Mrs. Steve Foreyt, Mr. and Mrs. Frank Kirch, Joseph Jerabek and daughter Veronica and son Joseph Jr., and Mrs. John A. Burich of Reedsville.

Mrs. John A. Burich was, of course, Anna. In all, fifteen adults helped husk corn into the wee hours of the morning. They were mostly relatives, except for the Foreyts and Jerabeks, who were nearby neighbors. Those families often traded work back and forth.

FIREWOOD REQUIRED THE SAME GANG

In the days before chain saws, hand saws and axes were needed to cut firewood to the right size. If the farm family didn't make firewood, they bought coal to heat their home. As farm machinery evolved, tractors came equipped with belt pulleys to propel circular saws for cutting firewood. It often took multiple men to do the work.

On February 19, 1945, six neighborhood men arrived at the Pritzl farm to make firewood. While the men worked, Julia and Anna were back in the kitchen. Just like the threshing crews, the men received two meals, beer and wages for the day.

No notation was made about Anna's pie selection that day. But one can bet those famous pies were on the dinner table.

NEIGHBORHOOD FUN

Many a neighborhood group got together for card nights. Sheepshead was the game of choice for many in the area.

Coin showers were also commonplace. Postcards and newspaper advertisements would often issue a grand invitation to join in a celebration prior to a couple's nuptials.

Here's an invitation for Julia and Elmer's niece, a daughter of Julia's sister Agnes, from May 12, 1953. All in the neighborhood came for a night of dancing and drinking for celebration.

Yourself, Family and Friends Are Cordially Invited to Attend the

Coin Shower

—— Given in Honor of ——

Miss Janice Kubsch

and

Mr. Bernard (Butch) Reif

May 23rd, Gosz's Ballroom, Polivka's Corners

Music from 9:00 till 1:00

Wedding Dance - Kubsch's Hall, Kellnersville

Tuesday, June 2nd, 1953   Music by The Goszlings

Coin showers were commonplace and a great opportunity to gather, celebrate and thank your neighbors for help given to one's farm family throughout the years. *Author's collection.*

Coin Shower
Given in Honor of Miss Janice Kubsch and Mr. Bernard (Butch) Reif
May 23rd, Gosz's Ballroom, Polivka's Corners
Music from 9:00 till 1:00

Wedding Dance
Kubsch's Hall, Kellnersville
Tuesday, June 2nd, 1953
Music by The Goszlings

Hunting was another neighborhood pastime. It was the reason that groups such as West Maple Grove, Rockea and other hunting clubs sprang up locally. On one occasion, Elmer made a rare entry in Julia's journal on the very matter.

"Today, Fuzzy Labitsky and I each shot a goose at Becker's Lake," he wrote of their Saturday, October 3, 1953 hunting excursion about three miles from their respective farms. "Fuzzy's Canadian Goose was 10-½ pounds and mine was 9 pounds."

One can read the joy emanating from that journal entry.

True to this day, even though farmers work extremely hard, they find a little time for fun with their neighbors.

6

THE COWS ARE OUT

I f you grew up on a dairy farm, your heart instantly starts racing and your pulse quickens whenever one proclaims, "The cows are out!"

That's an "all hands on deck" alert to collect the herd—a 911 emergency for immediate action. By their very nature, cows are herding animals. When one member of the herd wanders off through a compromised fence or an open gate, the entire herd will soon follow the adventurous bovine as if the Pied Pieper himself were magically leading the cows to a mystical land filled with even greener pastures. While that really isn't the case, a frolicking herd can decimate a newly planted field or cause irreversible harm to a crop nearing harvest. Plus, there's the concern that roaming cows on a roadway can cause an accident.

Fencing laws in the Badger State are quite unique since cows have long been the heartbeat of Wisconsin's economy. That being the case, Wisconsin State Statutes extend unique privileges to both cows and their owners. Chapter 90 of the Wisconsin State Statutes, originally written in 1875, sets forth the general rules regarding agriculture fences. It contains nearly six thousand words and remains the law of the Badger State to this day.

For starters, a well-maintained fence that complies with state statutes essentially "sets a farmer free" by limiting liability caused by cows and many other farm animals. This liability-reducing waiver does not apply to testosterone-infused males such as bulls, stallions, boars, rams and billy goats. These males require a much stronger fence.

On the flip side, a landowner who does not maintain or keep in good repair his or her part of the legal fence has brought the liability back to

Area farmers grazed cattle as often as possible because the alternative was harvesting and storing loose hay. It was a labor-intensive cropping activity; this photo captured Elmer's ninety-ninth load of the growing season. *Author's collection.*

themselves. Even if you don't own animals, you as a landowner may need to maintain a fence for your neighbor who owns cows. Yes, you may need to build a fence for your neighbor—more on that later.

By 1900, twenty-five years after Wisconsin's first fencing laws were inked, fences lined the rural landscape to keep cattle herded on each respective farm. Most farmstead driveways also had gates that swung shut to keep neighboring cows from strolling up driveways when herds were being moved to pastures down the road from unconnected farmsteads.

DESPERATE TIMES, DESPERATE MEASURES

When the Dirty 1930s cast their long, dark shadow over America, Wisconsin's lush green pastures began burning up and turning brown. That was a major problem, as the Badger State was home to over 2 million dairy cows—the most in all of America—in the early 1930s. That number eventually peaked at 2.37 million in December 1944.

The Dirty Thirties prompted the Burich family and other neighbors to create roundups to drive cattle down rural roadways to the nearby Collins Marsh in search of lush vegetation. Lacking access to the marsh, one local dairyman came up with another idea.

Graze the road ditches.

"Hanna Willie would just let his cows go," said Julia, calling the farmer by his nickname. Willie had a homestead just three forty-acre parcels, or three-quarters of a mile, to the south of Pritzl's Pine Haven Farms.

"He would let them pasture all along the road by our farm," Elmer said of the road that would eventually be known as Hickory Hills Road. "There wasn't much traffic back then."

"There wasn't much grass either," added Julia, recalling the desperate situation. "Those were dry years."

"Dry years is right," confirmed Elmer. "Very dry."

Knowing the plight of the financially strapped farmer, neighbors just looked the other way as Willie's cows strolled up and down the road eating the grass-lined ditches.

Generally, Willie's cows roamed on Sundays and were back on farm pastures during the weekdays as more automobiles started rolling down country roads.

Just Didn't Quit

Coming home from a Sunday church service and seeing your neighbor's cows grazing on your lawn wasn't a pleasant experience. Elmer hustled to change his clothes and chase the bovines off their farm. *Author's collection.*

When the rains returned in the early 1940s, Willie just kept grazing the ditches, as the practice had become second nature for him—plus the feed was free. However, Willie's ditch-grazing cows were wearing out their welcome to neighbors. For Elmer, the breaking point came late in the spring one Sunday morning.

"We came home from church, and Willie's cows were all around our house making a mess and ripping up our lawn," Elmer said, noting there were cow pies—manure—plopped all over the lawn.

With that, Elmer and Julia sprinted into the house, changed clothes and chased Willie's cows off their farmstead. After the bovines were dispelled from the property, Julia pulled the barbed wires across the driveway to close the wire gate. Elmer made his way to his neighbor's farm to discuss the matter.

The situation improved to some degree, as Willie quit grazing the ditches; however, his cows still got out from time to time because his fences were in terrible condition. The saying "the grass is always greener on the other side of the fence" has a direct meaning. Willie's cows knew the weak spot in the fence, and they knew that the grass *was* greener on the other side and escaped at every available opportunity.

I Will Hold You Liable

After those cows had ripped up their lawn earlier that spring, Elmer finally had enough of Willie's free-ranging cows. He went over to lay down the law—Wisconsin State Law Chapter 90.

"I planted winter wheat in the field by the road and I took the fence down. I will hold you responsible if your cows get out because I haven't got the time to put the fence back up.

"If they go in that wheat field, you are going to pay for it," Elmer said of the wheat crop.

After that conversation, Willie and Elmer never spoke much again. Willie was mad. But he fixed his fences.

Why? The law was on Elmer's side.

Since Willie's cows were repeatedly escaping from a poorly maintained fence that adjoined a township road, Willie was on his own to repair it. He replaced the old fence with a four-strand barbed-wire fence to meet codes detailed in Chapter 90.

With that new fence, Willie's cows stayed home.

Fences Are a Shared Expense

Wisconsin values its cows and livestock. To this very day, if one farmer wants to graze animals, the two adjacent property owners are each on the hook for 50 percent of the fencing expenses. That's even the case if the second landowner does not want to graze his or her property. So, in the previous example, had Elmer and Willie shared a property line, Elmer would have been on the hook for 50 percent of Willie's fence. Since cows were getting out on the roadway fence line, the cost was all on Willie.

How does the 50-50 cost-sharing get measured under Wisconsin statutes?

Generally speaking, if you, Neighbor A, stand in the middle of your property line and face Neighbor B, you have to maintain the half of the fence on the right and Neighbor B maintains the half on the left.

Chapter 90 also has extensive language to settle disputes. If a dispute must be settled, that matter generally falls on the township's supervisors, who become "fence viewers" under state law.

Woven wire fences must be fifty inches high with the bottom not over four inches off the ground. Shorter heights apply to barbed-wire and high-tensile fences and require four evenly spaced wires with the top wire being forty-eight inches above the ground. Electric fences are required to have two wires, with the top wire being between thirty-four and thirty-six inches above the ground.

It Goes on Your Tax Bill

If you are not a "good neighbor" when it comes to fence building, the state law is crystal clear. Towns and other municipalities are required to impose liens on landowners who fail to pitch in for the costs of maintaining or repairing shared partition fences that divide agricultural land. Failure to pay adds 1 percent interest every month until the bill is paid in full. And if the neighbor does not pay for the fence, the money comes from the township's treasury followed by an assessment to the noncompliant neighbor just as if it were another property tax bill.

As it turns out, this law also applies to cities.

The circuit court in Jefferson County ruled that the City of Watertown had to pay its fair share of the fencing costs for Stuart and Janet White, nearby farmers and adjacent landowners. The City of Watertown did not like the first court ruling and appealed to Wisconsin's District IV Court of Appeals. In 2017, that three-judge panel reaffirmed the lower court's ruling as it relates to the 1875 law.

"When qualifying land is in a city or village, that city or village must administer and enforce Chapter 90 the same as a town would if the land were in that town," wrote Judge Paul Lundsten for the three-judge panel.

Indeed, good fences keep cows inside their pastures. Good fences also help make good neighbors.

7

IT CRACKED LIKE LIGHTNING

ZZZZZZZZZZ pang!"
The sound thundered down the hill and reverberated throughout the cow barn.

"ZZZZZZZZZZ pang!"

A few minutes later, another lightning bolt sound rang out from the Niagara Escarpment ledge rock, known to the locals as "The Rock." The sound mimicked a thunderstorm in late August when the humidity levels approach 90 percent—the kind that booms and rattles you awake in the middle of the night.

But these big booms were taking place in early April. There wasn't a cloud in the sky, and the sound emanated out for months.

Booms for Over a Year

"I never heard anything like that in the woods," said Elmer.

"For the first few days, Elmer was jumping up and down every time a lightning bolt and thunderclap rang out," chuckled Julia years later.

However, Elmer and Julia weren't chuckling in the spring of 1952. The couple, married fourteen years earlier, was steaming mad—mad that those trees were getting cut down.

"Elmer didn't like the fact that the woods was being logged," said Julia of the land the couple had farmed since 1939. She wasn't the least bit pleased either. The couple didn't own the property, and it looked as if this spring would be the first time Elmer and Julia wouldn't work that land.

Those who knew Elmer knew these two things about him: he loved trees, and he loved woodworking. Elmer was being reminded that both of those passions were being stripped from him every time he heard "ZZZZZZZZZ pang!"

The sound continued for months due south of the cow barn door, and the subsequent percussions even shook the farm buildings.

THE THUNDER'S SOURCE

The thunder emanated from Grandma's Forty—as in Anna Burich's forty-acre property. Anna's father, Wencel Satorie, bought the property on November 24, 1876. The Bohemian immigrant paid off his mortgage less than three years later on February 25, 1879.

While Wencel controlled the land his entire life, he transferred it to his unmarried daughter, Anna Satorie, one year before she married John Burich. With that transaction, Anna became the first woman in Manitowoc County's Rockland Township to own land. She received the land for a dollar after agreeing to take care of her widowed father. Anna was deeded the land on July 11, 1905, and her father lived another sixteen years.

Anna held the Satorie Homestead until her death on April 5, 1951, with the area's plat books always indicating "Mrs. John A. Burich" as the owner of that 40-acre parcel that held its own separate distinction from the Burich Family Homestead. Hence the family later called it "Grandma's Forty." The nomenclature sticks to this day.

Instead of willing it to Julia, her farming child who was already caring for the main farm, Anna left it to her four living daughters to deal with divvying up her beloved homestead.

Elmer and Julia had purchased the Burich Homestead over a decade earlier following the passing of Julia's father, John. As part of the purchase agreement, Anna had the right to live in the family's house for her "natural life on earth," per the exact words in the land abstract.

All was well for the first decade of cohabitation between Elmer, Julia and Anna.

Then Julia and Elmer wanted to make a few changes to the thirty-year-old home—the house that Anna and John had pinched pennies to build in 1916. As far as Anna was concerned, the home was perfect.

"Today we have started on our remodel and plumbing project," wrote Julia in May 1949, just four months after the birth of their third child, Rosalie. "Elmer and I tore the closet out, and the kids helped us lathe the walls," she continued, noting the preparation process for plastering walls with children Jacque and Elmer John now helping do some work. Fred and Elmer Krueger were the plumbers.

"Mother went on the warpath the minute the plumbers came," continued Julia. "Just because we moved her old plumbing fixtures, she won't use the upstairs bath!"

That didn't stop Elmer and Julia. The plumbers worked all of June 9 and continued June 13 to June 18. That was two men for eight to nine hours a day.

Anna Burich refused to talk to anyone.

She and her husband, John, toiled on building that house for three years. As far as Anna was concerned, there was no need for changes. She was not just hurt; one might even say wounded.

TEMPERS BOILED AGAIN

With the plumbing completed, Elmer and Julia hired Emil Svatek to make new kitchen cabinets.

"Mom pulled a tantrum about the doorway kitchen...so did Elmer," wrote Julia among the tension in the three-generation family all living under one roof.

"I don't like it either, but whatcha gonna do?" Julia asked herself.

Julia clearly wanted new cabinets and was willing to do away with the kitchen door that split the workspace. That doorway made food preparation and cooking a challenge. The entire project had a financial cost of $329.97, but the collateral relationship damage was much more.

After having a smooth transition on the purchase of two hundred acres a decade earlier, Anna wouldn't even talk about selling her last forty acres. That topic was now off the table.

IT ENDED IN A FLASH

On Thursday April 5, 1951, less than two weeks after celebrating Easter, Anna's life ended abruptly. She died at the kitchen table while eating breakfast with Elmer and Julia—the same table where Anna had drafted the sale agreement transferring the two-hundred-acre farm homesteaded by the Burich family to Elmer and Julia.

Anna suffered a massive stroke.

Elmer jumped into the car and sped to Reedsville, seeking out both Dr. E.C. Cary and Reverend Bill Koutnik. On arriving at the family homestead, Doc Cary said there was nothing he could do. Father Koutnik delivered Catholic Last Rites as Anna would have wanted.

The family matriarch had died.

While they had their differences in the final years, Julia and Elmer were crushed. As they agreed when buying the farm, Elmer and Julia covered every funeral expense. True to form, Julia recorded every detail, including the $485 casket and vault, $30 for grave digging and $28 for flowers. Doc Cary received $10 and Father Koutnik $11. The Cemetery Association received $25, and the stone grave marker cost $35.

"We served our boiled home-cured ham. Delicious!" wrote Julia in her diary. "Our own roast beef that I roasted, turned out beautiful. I made the gravy and Agnes [sister], and I peeled the potatoes," continued Julia. "Leona Burich [neighbor] made lemon pies. Beatie [sister] brought apple pies. Sally Vondrachek [neighbor] brought a cake.

"I served date cake, kolaches, rolls, cabbage salad, peas and carrots, and a fruit salad," wrote Anna's youngest daughter, Julia.

"I was a nervous wreck all throughout and after, too."

Sisters Came on the Scene

Why the nerves? Julia's older sisters would be coming over more often to divide up their mother's possessions.

This can be a challenging process for many families. Since Anna Burich lived in the same home from 1916 to 1951, those decisions would all be made at Anna's last residence—the home now owned by Elmer and Julia for the past twelve years.

"The girls are supposed to be dividing all Mom's personal things," wrote Julia, who was one of those "girls."

"Agnes took only nice things. Mary took most anything. Beatie took some nicer items too. I simply got what I gave her as gifts," wrote Julia, who appeared to feel slighted.

In her mind, that would be the least of her slights. Then there was the land known as Grandma's Forty.

Elmer and Julia had been farming it since 1939. Two of Julia's oldest sisters received $4,000 at the time of their weddings as a gift from their parents, Anna and John Burich. With the Great Depression's smothering effect, Beatie received only $1,000 and two cows.

Given Julia's father died less than a year after her wedding, Julia received no money. She had assumed that Grandma's Forty would become her land.

Julia's opinion didn't matter at that point. A vote took place. The vote count was 3 to 1 with Julia being the lone dissenter. The Satorie Homestead would be sold.

Presumably, Elmer and Julia had a chance to buy it. However, the asking price was likely too high in their minds. That's because of all the standing timber on the property that had grown mighty tall and wide since Wencel Satorie bought it in 1876.

As executor, Mary Kalies, the oldest of the sisters, went before Manitowoc County judge Jerome V. Ledvina. He ordered that the forty acres be split into

In a photo taken after settling their mother's estate, sisters Beatrice, Agnes and Mary (*standing*) have smiles on their faces, as their payday was right around the corner. Each was set to receive $1,750 for her share of their mother's homestead. Julia's grim facial reaction (*seated*) tells another story—one of sadness—as she could no longer farm her mother's Satorie homestead. *Author's collection.*

four equal shares between Anna Burich's four living daughters: Mary Kalies, Agnes Kubsch, Beatrice Sleger and Julia Pritzl. The decree was dated and signed on November 7, 1951.

Just days earlier, Mary had hatched a plan to sell the property to E.A. Quinnette and M.J. Saenger. Those two men owned a nearby lumber company, and they wanted the valuable oak and maple logs. Quinnette and Saenger paid $7,000 to obtain the property.

Sisters Mary, Agnes and Beatie quickly cashed the checks. That included the $200 Mary paid herself as administrator of the estate. A disgusted Elmer suggested to his wife, Julia, "Have them pay us in spring." The *them* were the lumbermen Quinnette and Saenger.

On the rest of the matter, Elmer held his tongue. While badly bruised, he was not ready to wave the white flag, as he had been through more trying times.

Elmer had always considered Anna Burich his mother, having lost his own mother as a teenager. He also had been farming the land as if it were his own. Given the situation, that would not happen in the spring of 1952, as Quinnette and Saenger now held the deed.

That didn't sit well with Elmer. He started working through the situation because he knew his mother-in-law's heart. Despite their recent differences over the home remodel, she had loved her father's homestead. Elmer thought in his heart of hearts he must somehow get Anna's homestead back.

8

HE HATCHED A PLAN

ZZZZZZZZZ pang!"

The reoccurring sound of logs being dropped caused Elmer to double down on his resolve to find a way to regain his mother-in-law's homestead. Elmer put on his thinking cap for days, weeks and then months. For him, that meant he rolled a cigarette using Top Tobacco paper and Velvet Pipe Tobacco.

During the winter of 1951–52, Elmer puffed on cigarettes as if he were a steam locomotive climbing the vaunted Sierra Nevada Mountain Range at the opening of the Transcontinental Railroad. As he puffed, he stared out the barn door looking to the south at his late mother-in-law's homestead.

With each puff, Elmer asked himself, "What can I do?"

Puff.

He was slowly developing his plan.

Puff.

Then one day, the steam whistle blew, and Elmer spoke his plan out loud to his wife, Julia.

"I want to buy Grandma's Forty," said Elmer.

"I don't care for it," Julia snapped back, not wanting to reopen wounds from months earlier when her three sisters voted to sell the property.

"We have enough work. Why would they even sell it?" Julia quipped back, with the *they* now being the lumbermen.

"I already talked to Quinnette," Elmer said of a conversation days earlier with the lumberman. Elmer had posed this question, "If you could have five years to log off the woods, what you would want for the property?"

Quinnette thought for a moment and said, "$3,000."

"If I give you $100 down, can I farm it this spring? I'll get you the remaining $1,150 in the fall. The final $1,750 would come from Julia's share of what you owe her for the sale," suggested Elmer to Quinnette.

"That will work for me," Quinnette responded.

As part of the deal, Elmer would be giving up $1,750 of Julia's inheritance—the money each sister received for their share of the forty-acre homestead. Elmer and Julia had deferred receipt of that money until the spring of 1952.

"I'll talk to my wife and get back to you in a few days," said Elmer. The men shook hands, and Elmer walked away with a smile on his face. Now if he could convince Julia that this was a good idea, Elmer could be at peace, as he would have just reconciled the entire family homestead.

Julia Wrote and Wrote

She loved her husband, and even though the scars were not healed from the battle with her sisters, Julia warmed to the idea that her husband had put so much thought into buying back her mother's homestead.

"So, I suppose we can pay that much for it?" she asked Elmer.

"I wonder if the logs are worth $4,000?" Julia later asked herself in her diary.

"I don't suppose it's a not bad investment for us. I'm sure it would make Mom happier to know Elmer is handling her pride and joy."

"We will buy it," she wrote in her journal.

On May 7, 1952, six months to the very day after the Burich sisters voted 3 to 1 to sell their mother's homestead, Elmer and Julia bought it right back.

Knowing the Pritzls, that date was intentionally selected. The following disclaimer was stamped into the land abstract:

> It is understood that the grantors (Quinnette and Saenger) reserve for themselves, their heirs and executors, and assigns the right to harvest standing timber which is not less than 11 inches in diameter at the stump and the right to harvest standing pine timber which is not less than 8 inches at the stump. The grantees (Pritzls) shall be entitled to the tops, culls, and buts which may result from the harvesting of the standing timber herein referred to.
>
> The grantors shall have the right of ingress and egress to the said real estate for the purpose of harvesting timber, but they shall in no event damage

growing crops, and in the event that they do, they shall compensate the grantees for such damage.

These rights shall be limited to May 7, 1957. (However, if the grantees wish to extend this period of time for an additional three years, they may do so.) The harvesting of standing timber hereinabove referred to shall also include windfall timber.

THE THUNDER ROLLED

Lumbermen Quinnette and Saenger cut everything that could be legally cut by the terms of the contract. Sugar maples and red oaks eleven inches and greater hit the rock.

"ZZZZZZZZZ pang!"

So did the pines greater than eight inches.

"ZZZZZZZZZ pang!"

"They pretty much clear cut and logged off the entire forty-acre parcel," said Julia. "You could hear those logs crack on that rock all the way to the farm," she added.

"Oh, they were big. Oh boy were they big," exclaimed Elmer of those logs.

"ZZZZZZZZZ pang!"

"Even though we now owned the property, Elmer jumped up and down every time a tree hit the ground or whenever another truck load of logs was hauled out," said Julia.

For Elmer, it was almost as if family members were dying every time another tree hit the rock outcroppings.

"ZZZZZZZZZ pang!"

A GOOD BUY?

"We bought it!"

"Period!" wrote Julia in a 1953 journal entry.

"It panned out!"

"Harvest 1952 finished up on Ma's before August 2. We threshed 109 bags of barley up there," she continued, sketching out the math of $3.70 a bag for a $403.30 sale.

"The rest is in buckwheat," she wrote, noting that the couple planted the exact crop her grandfather Wencel Satorie planted seventy-five years earlier on the homestead.

By the time 1960 came on the calendar, Julia had become convinced that buying her mother's homestead back was a good idea. Elmer and Julia are shown here on their twenty-second wedding anniversary. *Author's collection.*

HEAT SEALED THE DEAL

Eight years later, Julia wrote her final thoughts on the decision to buy her mother's homestead.

"March 1960: We've used the maple wood from Grandma's Forty up to now," said the penny-pinching purebred Bohemian. "So, we saved a lot on coal—not one ounce did we buy. So that's eight years at $200 per year for coal. We got $1,600 back on coal savings. Winter 1961, we had to buy 1,800 pounds of coal. Had good crops up there, too."

Between the $403 from the sale of the very first barley crop and the $1,600 saved by burning wood instead of coal, Julia had her inheritance money back, even though she never penned that final statement in her journal.

With that last journal entry on the subject, the lightning storm was over.

Elmer went about planting trees to heal the scalped land. More importantly, the family farm was whole again.

9

THE $63 BABY

B y 1950, World War II was in the rearview mirror and America had entered its baby boom. War veterans were getting married to their sweethearts, and the babies soon followed the nuptials. Booms of this nature are typically inspired by prosperity and propelled by a bright economic outlook. Postwar America had all those metrics covered in spades.

In early January 1949, Julia and Elmer Pritzl made their way to Manitowoc's Holy Family Hospital. On January 16, Julia was assigned Room 413. The very next day, their third child was born: Rosalie Mary.

By medical standards, childbirth had come a long way in just one generation. Both of baby Rosalie's parents were born at home: Julia in the family's farmhouse and Elmer in the family's home in nearby Brillion.

There was good reason that expecting mothers in the 1950s made their way to modern-day hospitals. If there was a medical emergency, both newborn and mother had a far better chance of survival there. For Julia's mother, Anna, both of her baby boys died shortly after birth. Thankfully, Anna pulled through those difficult births, as Julia was literally the baby or the last born of the family. Had Anna died in childbirth, Julia would never have been born.

Then there was the matter of better records when one was born in a hospital. Although she was born on October 30, 1918, Julia's parents, Anna and John Burich, didn't file her birth certificate until December 11 of that year.

As for her husband, Elmer, he never had a birth certificate as a child. While there was a record of baptism filed on August 15, 1917, at Brillion's St. Mary's Church, Elmer Wilfred Pritzl didn't obtain his birth certificate until later as an adult. That was only after others wrote letters on his behalf and Elmer took them before a judge to obtain a birth certificate to confirm that he was indeed an American citizen.

The ultimate irony, though, is that Uncle Sam had no problem taking Elmer's signature years earlier when President Roosevelt initiated the Selective Service and Training Act that required all young men to sign up for the military draft.

An Eight-Day Stay

By the time baby Rosalie came on the scene, hospital stays for delivering a baby were quite an affair. Julia checked in on January 16 and checked out on January 24. Coincidentally, one year later to the very checkout day, Rosalie's future husband, Randy, was born in Green Bay.

Julia kept the hospital bill and verification of payment that totaled $63.43. The total included:

$36.00 Room rate @ $4.50 per day
$5.00   Anesthesia
$10.00 Delivery room
$8.00   Care of baby
$0.75   Medicine
$2.50   Dressings
$1.00   Laboratory fee
$0.18   Telephone calls

While the $63.43 bill for an eight-day hospital stay sounds like a financial steal by today's standards, remember that the average annual income for an American family in those days was $3,300 according to the United States Census Bureau. Price increases over time, known as inflation, change one's perspective.

Even so, that hospital bill was a deal.

In 2015, the United States Department of Agriculture (USDA) estimated that it cost $233,610 to raise a newborn to adulthood. That cost estimate doesn't include a technical school or college education.

When Rosalie Mary Geiger (née Pritzl) was born in January 1949, the hospital bill for the eight-day stay was $63.43. *Author's collection.*

That $233,000 figure is a national average. The urban Midwest has a lower total at $227,400, while rural America was $193,020 in 2015.

Of the cost categories, housing topped the budget list at 29 percent. That's followed by food, 18 percent; childcare and education, 16 percent; transportation, 15 percent; healthcare, 9 percent; and clothing, 6 percent. Then there's the all-encompassing category of miscellaneous that envelopes the remaining 7 percent.

While costs vary over time, only one category shot through the roof from 1960 to 2010. That's medical care—it exploded by nearly 140 percent according to government data.

ANNA LIVED ON

Crushed by the loss of the lady who gave them so much, Elmer and Julia bestowed the highest praise possible to Anna Burich. When their fourth child entered the world in the spring of 1952, the couple named her Annette—the modern-day rendition of Anna. As it turns out, little Annie filled the farmstead home with the same joy that Anna Burich had all her life.

By the time baby Annette happened on the scene, costs were already climbing for maternity care. Julia was again assigned to Room 413 at Manitowoc's Holy Family Hospital. This time around, the room rate jumped to $5.50 per day, and the entire six-day stay cost Elmer and Julia $84.13. That bill also was marked paid on the very day Julia was discharged by the medical staff.

The total included:

$33.00 Room rate @ $5.50 per day
$3.00   Anesthesia
$10.00 Delivery room
$12.50 Care of baby
$9.45   Medicine
$3.50   Dressings

$5.00   Laboratory tests (Baby)
$5.00   X-Ray Service (Baby)
$2.00   Oxygen (Baby)
$0.68   Telephone calls

In just three years, medical costs climbed significantly as detailed by those two delivery bills. Now, to be fair, Annette was a "breech" baby, so there were some extra costs.

For starters, in the process of turning the newborn in Julia's womb, the baby's arm was broken, hence the X-ray. Despite all the extra medical attention for that baby, the hospitals were starting to reduce maternity stays; baby Rosalie had an eight-day stay, while baby Annette had a six-day stay.

A Look Back to a "War Baby"

Elmer and Julia were busy supplying Uncle Sam with clover seed from their two-hundred-plus-acre farm during the height of World War II. The oil from that crushed seed helped lubricate the cannons on naval vessels and was the reason Uncle Sam kept Elmer at home on the farm even though he was of prime military age. The War Department valued that clover seed as a higher priority than having Elmer in a military uniform.

It's also the reason Elmer and Julia were able to have a baby boy born to them on February 19, 1943.

"Elmer Jr. born February 19th at St. Vincent's Hospital. February 22 he was circumcised, and the total hospital stay cost $10," wrote Julia, who didn't even receive a bill for that hospital visit.

There you have it—medical costs boomed right along with the baby boom. A 1943 hospital stay to have a baby cost $10; in 1950, $63; and in 1952, $84.

From 1943 to 1950, the medical costs jumped 530 percent (although price caps were in place due to the war). From 1950 to 1952, costs jumped another 33 percent for the same parents at the same medical center. Babies were indeed booming, and babies were about to become big business across America.

10

WISCONSIN'S HONEY CAPITAL

Reedsville Today: in the land of cows and clover."
That was the headline of the 1921 business directory for the village
that had a bustling population of some six hundred people.

"Situated in one of the finest dairy and agricultural regions of the state, the soil is of a high grade clay loam variety and especially adapted to dairying, livestock raising, and diversified farming," the directory stated, as recounted in the book *History of Reedsville to 1976.*

"Dairying, livestock raising, clover seed growing, and honey are the chief industries which are very extensively practiced."

Honey.

Just how much honey did the Reedsville area produce in the 1920s?

While alfalfa was on its way to becoming the queen of forages, clover had a strong foothold in the area. Although bees like alfalfa, they love clover.

Looking at our family's crop records in the 1930s, 81 percent of their farm was either planted in alsike clover or sweet clover. And if you were making dry hay in those days, you let the clover go to seed because the crop would dry better. Those seed heads also became homing beacons for bees. Like the Burichs, most farmers in the area grew clover.

John and Anna Burich did not have any beehives. However, Frank W. Kirch had hives—and plenty of them. Frank was married to John's sister Elizabeth. After she died in 1923, John and Frank continued to partner on the honey enterprise.

NORTHEASTERN WISCONSIN BEEKEEPERS

Reedsville area residents were active members in the Northeastern Wisconsin Beekeepers Association. There was even a three-day bee school held in Reedsville that Edward Hassinger Jr. wrote about in the April 1920 issue of *Wisconsin Horticulture*: "The Association instructed the Board of Directors to appoint a competent person as grader for the Association. The place for grading will be decided upon by the Board of Directors."

TWO HUNDRED ITALIAN QUEENS

"The members of the Association also ordered 200 purebred Italian queens. This is one of the most important steps taken in recent years by our local beekeepers and to my mind is the beginning of what will soon be a state-wide selling organization," Hassinger's article noted.

"Snow nearly gone. Outdoor bees had a few more days warm enough to fly and contrary to expectations have wintered fine with just a few exceptions. It seems that many hives had a few bees filled with feces early in the winter and when those had left the hives, the balance of the bees in each colony were O.K.," concluded Hassinger. So, what became of the fledgling effort? The buzz grew louder!

An April 8, 1921 report, as published in the May 1921 issue of *Wisconsin Horticulture*, sheds more light on the matter.

"April 8—Condition of bees 100 percent better than last year," wrote Martin Krueger, reporter for the Northeastern Wisconsin Beekeepers Association, which now had 51 members.

"Winter losses about 5 percent due to queenlessness. Clover was thought to have been killed this winter but looking the fields over it seems to be fairly well. About 25,000 pounds extracted honey on hand," Krueger continued. Now 25,000 pounds of honey sounds like a big number.

It's been estimated that an average hive of bees can make 20 to 60 pounds of honey in a season. That means that on the high end, the Reedsville area had 1,250 hives. On the low end, it had 420 hives. In either case, that's a lot of bees.

By the time Reedsville's July 4, 1921 parade took place, the A.H. Rusch & Son Company had transformed itself from a lumber mill into a global beekeeping supply company. *Village of Reedsville.*

A NEW BUSINESS BLOSSOMED

A.H. Rusch purchased the lumber mill in Reedsville from Joe Dumas in 1893. Ironically, the building had been built by the Ludwig Rusch family in 1860. For many years, the mill was Reedsville's main source of employment. Of course, that was because the area had ample supplies of timber.

At one time, thirty-three men were employed logging during winter and sawing shingles during summer.

The last lumber was sawed in 1922. That year, the Rusch family did a business pivot and began to manufacture *Bee Keepers' Supplies.* Those three words have italic letters because that was the name of their new business catalogue. The business grew so much that catalogues were sent to approximately nine thousand beekeepers, shipping products worldwide as far as China!

The Ruschs were not only smart, industrious people but also leaders who gave back to the local community. A.H. Rusch, who eventually started the bee business, was Reedsville's first fire chief. He also served several terms on the village board and went on to be a founding member of the Reedsville Betterment Association.

SONS FLOURISHED

In 1930, the senior Rusch retired from day-to-day operations. That's the same year that his son Reuben bought half the business, joining his older brother, Albert F., in the family enterprise.

Like his father, Albert F. Rusch laid a wide swath in Reedsville. The younger Albert would serve as village trustee for two years and village president for seven years. He also was a charter member and first president of the Reedsville Chamber of Commerce and later held the office of secretary. Further, he was a charter member of the Zion Evangelical United Brethren Church, holding different offices in the church and Sunday School. He was elected trustee of the Wisconsin Conference in 1953. Rusch also organized Reedsville High School's first basketball team in 1905.

Rueben was fifteen years younger than his brother Albert F. However, he, too, was a leader. An avid sports fan, he was a member of various local basketball and baseball teams, particularly the Manitowoc County Basketball Team. One of the highlights of his vocations was assisting the coach of the 1946 Reedsville High School basketball team that won the Wisconsin state basketball championship in an era when all schools competed in one division. The Eau Claire team Reedsville defeated in the title game had more teachers than Reedsville had students.

Rueben eventually served as chief of Reedsville's fire department, village trustee and president of Reedsville's Methodist Church, and he was a Sunday school teacher for twenty-seven years. All three men are interred at the Zion Evergreen Cemetery southeast of Reedsville.

If you take a stroll through that cemetery, you will find these local businessmen, along with others with a passion for beekeeping, as identified by the beehives carved into the many granite headstones.

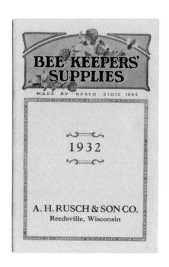

The 1932 *Bee Keepers' Supplies* catalogue from A.H. Rusch & Son Company featured everything one would need to operate an apiary. *Author's collection.*

After forty-three years of business, Reuben retired and terminated the enterprise in 1973. The office on Fifth and Menasha was sold to attorney Thuermer. The one warehouse on Fifth and Mill was purchased by the H.K.

Molding Company. Four additional warehouses were razed along with portions of the original lumber mill.

Bee Keepers' Supplies lives on in the world of eBay. That's where one can find a catalogue from the A.H. Rusch & Son Company of Reedsville, Wisconsin. The 1932 edition features forty pages that include everything one would need for an apiary career. Of course, an apiary is a place where bees are kept.

OUR GUARANTEE

"We absolutely guarantee our goods to be perfectly manufactured of the best material for the purpose. On examination, if our goods are not well represented, we do not ask you to keep them. Return same at our expense and we will refund your money, including any transportation charges you have paid," wrote the Ruschs in their catalogues.

"We have endeavored to list all of the principal articles necessary for use in the apiary, but if you desire something which is not listed, please write us, as we can furnish it as low a price as anyone," the Ruschs added.

"If you want some special hives, frames, etc., send us a list and sample or dimensions of same and we will quote you prices. We are in position to make anything in the line of beekeepers' supplies, if made of wood." Now that's something to buzz about.

11

WILD BEES COME TO ME

A land flowing with milk and honey."
 There are over twenty verses in the Holy Bible that reference a prosperous land "flowing with milk and honey," which God had promised to the Israelites and His children.

While the Badger State is located far away from the Middle East and the promised land known as Israel, Northeast Wisconsin mirrors the land "flowing with milk and honey." After all, Wisconsin had become America's Dairyland decades earlier, as it boasted a collective dairy herd with over two million cows shortly after World War II.

To be certain, the milk flowed, and so did curds that became cheese. So much cheese was produced in Wisconsin that the state's cheese factories were exporting 85 percent of its product outside its borders by the 1950s. Those cheese exports have since pushed past 90 percent. There was indeed a bounty to eat.

The area also had lush, rolling fields with alfalfa and clover to feed all those cows. That made beekeeping a natural endeavor, especially since some of those fields needed to be pollinated so the seed from the alfalfa and clover could be collected and sold to plant future crops. Hence the honey also flowed from the region.

Bees, Please Go Away

However, Daniel Geiger Sr. and his family were not wanting anything to do with bees and their honey on a hot early summer day in the early 1960s. They were far more concerned about turning over the cut alfalfa hay in the family's farm fields with the tractor and side delivery rake. By side raking the hay, the hay's wet underside would be flipped over to complete the drying process. Then, the legume and grass mix could be baled that afternoon or the next day. Once baled, it could be stored to feed the family's growing dairy herd during the fall, winter and early spring.

What had started out as a normal side raking routine of going up and down the cut rows of hay wasn't going as normal that day. Heck, the very question as to completing the task seemed in doubt.

As it turns out, one of the Geiger boys left the tractor and side rake in the field overnight. The young man had planned to walk back out to the field after the cows were milked in the morning and finish raking the hay.

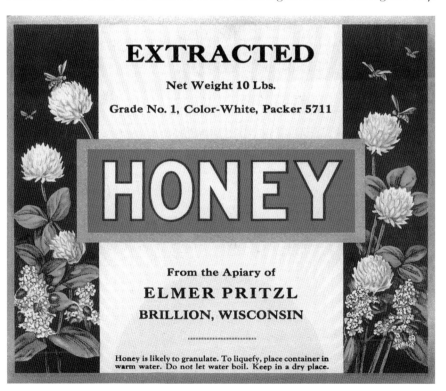

This label for saleable honey depicts the area's attributes, with clover being the nectar source for Elmer Pritzl's honey. *Author's collection.*

When he came back to the tractor and side rake that early August morning, though, he didn't go anywhere near either piece of equipment.

Bees were flying everywhere. They looked plum mad and even angry.

Sometime between the late evening and early morning hours, a giant swarm of bees had arrived and decided a nearby fence post on the property line would be a good place to set up camp. Not only was the fence post now covered in bees, but the tractor and side rake were as well.

The would-be side-raker scurried back to the home farm, not wanting to be stung.

Words flowed like a river of honey from the frantic young farmer as he shared the story with his grandfather Daniel Geiger Sr. Dan, as he was known to area farmers, didn't look that concerned. He almost had a "seen-it-before" look on his face.

"Where's the tractor?" he simply asked.

The field's location was divulged with one swift breath from the younger Geiger, now burgeoning with adrenaline. Dan hopped into the farm truck and drove to the scene.

Sure enough, there were bees everywhere. Dan quickly realized those bees weren't mad or angry. They were honeybees, and the swarm was just looking for a location to create a new home.

A lifelong farmer with tons of rural know-how, Dan knew this swarm had broken away from a nearby colony. He returned home and went straight into the farmhouse.

"Ida, can you give Elmer Pritzl a call?" Dan asked his wife. "Tell him we have a swarm of honeybees near our tractor and side rake. Ask him if he could come over at once and help us out so we can try to bale hay today."

Ida got on the phone and talked to the telephone operator, who transferred Ida over to the Rockland telephone exchange. She had a brief conversation with Elmer's wife, Julia, who said the message would be relayed to her husband right away.

"What would Elmer Pritzl know about getting rid of bees?" the younger Geiger lad thought to himself after watching his grandfather make the call.

Dan was like most farmers of the day. In the days long before the internet, the rural network of farmers knew one another's specialty. That network still runs as strong as any Google search. Dan knew Elmer was not only a dairy farmer but also beekeeper, as he grew fields of clover for seed. Elmer needed bees to pollinate his clover crop so he could sell that seed.

Even though Elmer lived in rural Reedsville, Dan and Elmer's bloodlines traced back to the same rural Kasson church dating back to the American

Civil War. The original Pritzl homestead could even be seen from the Geigers' farm. Even though they seldom communicated, they certainly knew of each other, as the local Kasson cemetery had a multitude of headstones bearing the names "Geiger" and "Pritzl."

LOOKING FOR A QUEEN

Elmer gathered his gear, including some empty beehives, and hustled over to the Geiger homestead. The hustle part was because a swarm of "wild" honeybees were a beekeeper's delight. Although it may have been a swarm, there was nothing wild about those bees. As a longtime beekeeper, Elmer knew those bees were a breakaway colony. That swarm represented local bees. Since they were local, they had survived the winter and safely made it to spring.

Those bees had good genetics. Elmer wanted those bees in his apiary, or colony.

Also, when the swarm leaves the hive for good, every worker bee carries as much honey as possible. That fence post had become base camp while scout bees were in search of a new hive. All Elmer had to do was provide that hive.

Elmer rolled up on the scene in his 1952 Chevy truck. He put on his beekeeper's suit, then his hat and veil. Lastly, he put on gloves. He calmly went over to the swarm as a small group of human onlookers, mostly young Geiger boys, looked on. Just like the bees swarming that fence post, the Kasson era had plenty of young Geigers in the 1950s.

Elmer kind of looked like he was an astronaut in his full white suit from head to toe.

First, he used his smoker to subdue the bees. The cool smoke was like a bee's lullaby. Between the smoke and the fact that the worker bees were full of honey, Elmer knew he wasn't in for much trouble that day. But the wide-eyed Geiger boys weren't that sure. Bees were crawling all over Elmer.

Elmer started to carefully sift through the bees. He was looking for something. Whatever he was looking for, he apparently found "it" and placed "it" inside the hive.

Then, he carefully brushed some bees off his outfit and walked back toward the road. After taking off his hat, veil and the top of the beekeeper's suit, Elmer rolled a cigarette and lighted it. As he puffed away, he started talking to Dan as the Geiger boys, mostly an assembly of grandsons, intently listened in.

A typical beekeeping scene from the 1950s and 1960s, this beekeeper was subduing the bees with cool smoke before working with the hive. *Author's collection.*

The boys stood there trying to figure out Elmer's game plan. As the men talked, the bees started to slowly migrate to the hive Elmer placed on the ground.

"Don't worry, those bees aren't that wild. They'll go crawling into my hive," said Elmer. "They're just looking for a home."

Elmer then talked a bit more about his apiary. "My bees arrived on June 13. At first, I had to feed them fifty-five pounds of sugar…five pounds to a hive," he said, explaining that these purchased bees didn't have a store of honey. "By July 31, the bees were really making honey. It's a steady mmmm-hmmmm," he said with a smile.

As for the swarm of bees on the Geiger homestead, sure enough, after about an hour, nearly all the bees had crawled into the hive. Elmer then went back into the field, picked up the hive and placed it in his pickup truck. Elmer was happy; he just received some free bees.

The Geigers were happy because those bees were gone from the tractor and side rake.

And as it would turn out, one of Dan's grandsons would be happy years later, too. Randy—one of the teenage boys there that day—had just met his future father-in-law, Elmer. Of course, Randy didn't know it yet. Randy would have to wait several years before he would meet his future girlfriend Rosalie. That would take place when Randy's first cousin Deanna Dvorachek, whose mother was born Hilaria Geiger, was out for a ride with Rosalie Pritzl. The teenage girls had been intending to go out horseback riding when Deanna drove by Grandpa Dan Geiger's home.

"Who is that?" Rosalie asked pointing to a strapping young man, well over six feet tall, who was talking to his grandfather. "That's my cousin Randy. Do you want to meet him? He's home for the summer from the Sacred Heart Catholic Seminary." So, the two teenage girls delayed their horseback riding adventure that summer day.

Indeed, the Geigers and Pritzls would once again gather in the late 1960s. However, instead of buzz coming from a swarm of bees, the buzz was emanating from a wedding ceremony and reception.

12

THESE LITTLE PIGGIES WENT TO MARKET

We sold 21 little pigs for $123," Julia entered into her ledger.
That would have been a rather common entry for a Wisconsin
dairy farm family in the 1940s. Julia's specific entry took place
on September 26, 1945. That same month, Emperor Hirohito signed an
unconditional surrender of Imperial Japan to the Allied Forces.

Back in America's Dairyland, pigs were nearly as valuable as cows. That's
because pigs were considered "mortgage lifters." As omnivores, pigs would
eat just about anything and turn those eats into cold cash. And the better the
eats, the better the pigs.

Pigs literally eat anything. Grain. Vegetables, plant roots and grubs, table
scraps, milk—and even whey—were all fair game.

In those days, whey was a near useless byproduct from cheesemaking.
Since Wisconsin was the world's cheesemaking capital, its cheese factories
churned out whey like Mississippi River water entering the Gulf of Mexico.

When it came to daily farm life in America's Dairyland, dairy farmers
would haul cans full of milk to the local cheese factory and bring home
whey. Each milk can holds roughly ten gallons or nearly ninety pounds.
Cheesemakers begged farmers to take the streams of "whey water" off their
hands. Most involved thought farmers were performing an act of mercy on
behalf of their cheesemaking cousins.

However, Upper Midwest dairy farmers, dating back to their pioneer
brethren, had been noticing this biological fact—little pigs fed whey grew
a whole lot faster than little pigs fed everything else. While reported science

hadn't completely caught up with the facts of Mother Nature, dairy farmers had figured it out and were raising piglets that would make Paul Bunyan puff out his chest with pride.

As it turns out, whey was packed with protein—the most complete amino acid profile known to humanity. The little pigs guzzled up the warm whey straight from the neighborhood cheese vats and were being fortified with the best building blocks to build muscle.

Today, an entire sector of the global dairy industry focuses on whey protein research. Dairy processors now can evaporate the water off the whey stream and capture that precious protein. Class III milk prices paid to dairy farmers in the Federal Milk Marketing Order system are based on cheese and whey. In countries like Vietnam, mothers will work for up to two weeks to purchase just one can of infant formula chock-full of whey proteins so that their young toddlers can grow up to be big, strong adults.

Whey has come a long way from being feed for pigs or as a fertilizer applied to fields as if it were manure from a pig or cow's rear end.

Reedsville's Fair Days

When Julia noted that those twenty-one pigs were sold on September 26, 1945, she called it the "September Fair." That last Wednesday of September may have been just another monthly event in Julia's eyes, but these days, people would say, "September Fair?"

The little village of Reedsville had a monthly fair? More on that matter can be found in the book *History of Reedsville to 1976.*

"One important feature in the history of Reedsville was the monthly cattle fair, held on the last Wednesday of each month. This was a day of bustle and business—a gala day," wrote the book's authors of Reedsville's early days. "Crowds by the thousands came from far and near, all roads leading to town for miles were lined with pedestrians, vehicles of all descriptions and livestock.

"Booths lined the main streets which were rented to transient merchants who offered for sale clothing, boot and shoes, pots and kettles, wagon loads of melons and fruits. Patent medicine peddlers, gamblers and loose fingered gentry were there…horse jockeys, gypsies, also."

Could that account be true?

Keep in mind, the little village of Reedsville had eleven taverns in those days—somebody was keeping those bar keepers in business in a place with a population of just six hundred.

This Reedsville street scene depicts the view just prior to its monthly fair. *Village of Reedsville.*

With a crowd of people coming to the village each month for its monthly fair, the Reedsville Motor Car Company sold not only Buick automobiles but McCormick-Deering seeders and disks as well. *Village of Reedsville.*

And when it came to little pigs brought to those fairs—some light-footed escapees reportedly scampered throughout Reedsville for days and sometimes weeks.

But for Julia and her husband, Elmer, those monthly fairs were an outlet to sell pigs just as they had been decades before for John, Albert and Thomas Burich. Those three men were Julia's father, grandfather and great-grandfather, respectively.

MARKETS WERE A CHANGIN'

As the world rebounded from the depths of World War II, food pipelines began to fill. Farmers no longer had an easy time selling little pigs, and Reedsville's robust cattle fair began to fade into history.

However, little pigs were still everywhere on nearby farms, and they needed a home. In 1956, there were thirty-five pig fairs in the state, accounting for 22 percent of the sales, reported the University of Wisconsin's College of Agriculture. Other sales outlets included direct sales to dealers at 38 percent and sales directly to other farmers at 30 percent. A variety of other outlets filled the remaining 10 percent of sales.

Those selling pigs were becoming disgruntled. Criticism centered around the lack of price discovery, crowded conditions and in some cases a lack of enough buyers at the once flourishing pig fairs.

That's when a few industrious farm folks came up with the idea of forming a co-op. By all reports, Norval Dvorak was among the ringleaders. In fact, he became the first manager for the Wisconsin Feeder Pig Marketing Cooperative. Julia's husband, Elmer, signed up as an early member in March 1958. Julia's father, John Burich, had long believed in cooperatives and was a charter member of the Reedsville Cooperative, which operates to this day as Country Visions Cooperative. Because of that, Elmer believed in co-ops, too.

"Feeder Pig Co-op Claims First!" proclaimed the headline in the April 1957 edition of *News for Farmer Cooperatives* published by USDA.

"This co-op believes itself to be unique in this country. It is the Wisconsin Feeder Pig Marketing Cooperative at Whitelaw, organized to cover eight northeast counties in the State. The co-op sells quality pigs for its 1,200 members at a premium with fieldmen of the co-op weighing and grading pigs on the farm.

Cooperatives gave farmers a voice in their business. Hence, Julia's father became a charter member of the Reedsville Co-op Association. The next three successive generations would belong to that co-op, too. This image was captured in 1936 by Art Neumeyer as he stood on U.S. Highway 10. *Manitowoc County Historical Society.*

"In its first 10 months, the co-op handled nearly 44,000 pigs for a gross income of $675,000. Four out of five pigs were sold outside the State."

Most of those pigs were heading to Iowa to be fattened up on cheap grain. Because of this co-op, Julia and Elmer were earning $15 per pig, not the meager $5 in that 1945 ledger entry.

FIVE TIMES THE GROWTH

Farmers flocked to the feeder pig co-op.

"Feeder pig production is big business in Wisconsin even though it is just a sideline to dairying," reported the April 1962 edition of the *News for Farmer Cooperatives*. "The Wisconsin Feeder Pig Marketing Cooperative, with headquarters at Francis Creek (moved from nearby Whitelaw), started its integrated production-marketing program in 1957 with 426 members and 3,000 sows. The co-op's volume has grown from 47,000 pigs in 1957 to over

200,000 in 1960. More than 4,000 members with its 28,000 brood sows in 58 Wisconsin and Minnesota counties are now producing pigs under a three-year contract with their co-op.

"Pigs weighing at least 40 pounds are graded and weighed on the farm and hauled to four assembly points by 16 part-time fieldmen. Here the pigs are sorted into uniform lots to fill specific orders and are shipped by contract carrier direct to buyers. Pigs are being shipped to most of the Corn Belt States and some to the east coast area. Some pigs go to the co-op's new display barn at Grundy Center, Iowa, where farmers can look at the pigs before buying. All other pigs are sold before they leave Wisconsin.

"In addition to marketing the pigs, the Wisconsin Feeder Pig Marketing Cooperative secures improved breeding stock, disseminates information on feeding and management, and provides field supervision of management problems. Producer-members agree to market through the association during the term of the contract all pigs sold by them under 100 pounds, with the exception of those sold for breeding purposes."

By becoming a member of the Wisconsin Feeder Pig Marketing Cooperative, Elmer and Julia Pritzl ended the family's four-generation run to the Reedsville Cattle Fair and turned over their pig marketing to the newly formed co-op.

Eggs also no longer headed to the Reedsville Cattle Fair but to the newly formed Lakeland Egg Cooperative in nearby Valders. The Pritzls received a $10.98 patronage allocation from Lakeland Egg Cooperative in June

The Wisconsin Feeder Pig Marketing Cooperative was the first of its kind in the country. Norval Dvorak, the first manager, grew the co-op's membership to four thousand strong in the first four years. *Author's collection.*

1955 and $5.87 from Land O'Lakes Creameries, which had a marketing agreement with the Valders-based cooperative.

That latter cooperative later dropped "Creameries" from its title and became known as simply Land O'Lakes—one of the world's most recognized cooperatives. By 1977, the egg co-op had merged into Land O'Lakes. That also shuttered the processing plant in Valders and resulted in the layoff of twenty workers.

When it came to merchandising dairy cattle, the Pritzls had turned that matter over to another cooperative—Equity Livestock—which had opened a location in Reedsville in September 1958. Indeed, Reedsville's cattle fair days were now over, but the little village would remain a cattle hub for decades to come due to Equity Livestock.

13

BYE, BYE, LITTLE BEES AND PIGGIES

These days, neighborhood watches are formally organized efforts so that local people can look out for one another. To be successful, these efforts require everyone's help.

While somewhat informal, the same situation played out in the countryside. However, everyone didn't always carry their weight.

The demise of Elmer's beehives came in the late 1960s when American foulbrood struck his hives. Foulbrood is a fatal bacterial disease caused by the spore-forming bacterium *Paenibacillus larvae*. The disease strikes strong and weak bee colonies alike by killing bees early in life. Without the next generation waiting in the wings, the hive eventually dies out.

There is no cure.

That means beekeepers who encounter American foulbrood must destroy their infected colonies, hives and all infected material to ward off its spread. It's the reason many states still issue apiary permits and track bee colonies.

It turns out that Elmer's neighbor never acted when his bees were infected with American foulbrood. Not only did that neighbor lose all his hives, but those honeybees also carried the disease to nearby fields. During pollination, those infected bees spread the disease throughout the neighborhood.

No more bees.

Elmer was so disgusted that he gave up beekeeping for good. That's because Elmer knew his bees, and he also knew American foulbrood bacteria could persist for decades in an area.

3998

THE STATE OF WISCONSIN — DEPARTMENT OF AGRICULTURE
Division of Plant Industry — Bee and Honey Section
448 West Washington Avenue, Madison 3

APIARY PERMIT

No. 323 - R

Under the authority conferred by section 94.76 of the statutes, this permit is issued to

Name FRANK W. KIRCH   Address Box 220, Reedsville, Wis.

and authorizes him to ~~move or sell~~ sell & move (8) colonies of bees, (20) hive bodies, (25) supers, (20) covers, (15) bottom boards, extractor, honey tank,

From Town of Rockland, Section 5 NE, Manitowoc Co. (Elmer Pritzl Farm)

to   VAN PATEK, RR1, Mishicot, Wisconsin
Town of Mishicot, Section 32, Manitowoc Co.

Date issued May 15, 1962   Void after June 15, 1962

DONALD N. McDOWELL
Director

John F. Long

Chief Apiary Inspector

Bees are susceptible to disease. That's why Wisconsin's Department of Agriculture later required permits to sell and move hives. In this application, Julia's uncle Frank Kirch moved his hives from Reedsville to Mishicot. *Author's collection.*

DON'T EVER GO IN MY BARN

Having learned dearly from the demise of his beehive, Elmer warned the local buyer for the Wisconsin Feeder Pig Marketing Co-op to stay out of his pig barn. Elmer always went into that barn by himself, and he hand delivered the forty-pound pigs to the local buyer. This went on for years without any incident.

These days, farming specialists would say Elmer was practicing good biosecurity.

Elmer was persistent on cleanliness, especially after his bees died. To his dying day, he would ask doctors, "When did you last wash your hands?" For Elmer, cleanliness was right next to Godliness.

One day, Elmer had gone to town, and the pig buyer stopped by the farm. That buyer violated Elmer's cardinal rule and went into Elmer's barn and plucked out the little pigs. Elmer knew it happened because the pigs were gone, and a check later came in the mail.

Weeks later, Elmer received another "payment"—erysipelas. It's a nasty disease that affects pigs. It nearly wiped out his entire hog operation. In little weaned pigs, high fevers and sudden deaths occur. It also affects adult hogs.

Unlike Elmer's bee plight, erysipelas can be cured with antibiotics. However, clearing the disease from the herd required repeated treatment of every animal with antibiotic injections plus a full scrub down of the facility. It cost Elmer both time and cash.

Elmer was so piping mad that he forbade that buyer from ever stepping foot on his property again. The buyer knew his physical well-being could be in danger if he crossed Elmer a second time.

These days, veterinarians and other advisors disinfect boots and equipment between farm visits. There's good reason, as it prevents disease transmission.

14

LET'S JOIN LAKE TO LAKE

In the spring of 1945, there was a simple need to find a can washer and a truck so a group of southern Manitowoc County farmers could continue to ship milk, since food safety regulations were becoming more stringent. Early in the journey, the dairymen confirmed another one of their long-held concerns—the region's milk prices were consistently under the state average.

The partnership that ensued among like-minded private farm business owners went on to develop and market the nation's first Grade AA cheese as certified by the United States Department of Agriculture (USDA). Perhaps more importantly, this startup business eventually reached $130 million in sales by its thirty-fifth year in business to become one of the country's premier dairy cooperatives. It all started when Henry Binversie and Melvin Lutzke paid a visit to Truman Torgerson, the county's newly assigned extension agent.

In the months leading up to that visit, the two Meeme township farmers, who served as president and secretary for the Hillpoint Cheese Cooperative, had tried to consolidate five similar-sized dairy processors in southern Manitowoc County and purchase a larger, privately owned factory. However, they learned an inspector threatened to condemn the plant if the group bought it.

There are likely two plausible reasons for the threat, with the first being the plant was not up to standards. The second, more likely, reason is that some folks were bringing out every stop to prevent the group of farmers from taking control of their own marketing destiny. That's when they turned to Torgerson, who came to Manitowoc County in 1944 after serving in a similar role in Rusk County for the two preceding years.

Upon learning the plight of these two farmers, along with the one hundred or so farmers they represented, Torgerson suggested the duo contact Rudy Froker, the highly respected dairy marketing specialist at the University of Wisconsin. A fact-finding mission ensued, and with Froker's help, the group learned that northeastern Wisconsin was trailing the state average pay price by eight cents per hundredweight. The hundredweight is the dairy industry's primary mechanism for pricing milk, and the 100 pounds of milk is equivalent to 8.3 gallons of milk.

Those eight cents were real money given that farmers were receiving about three dollars for every 100 pounds of milk. The reason for the difference was obvious—private capital failed to develop Grade A milk markets for producers. As a result, manufacturing prices remained low without Grade A competition. It also turned out that most of the area's cheesemakers liked it that way. Cheap milk meant more profits.

The remedy?

The area's dairy farmers needed to develop a Grade A market for their own milk. That would mean forming a cooperative and raising capital.

MEMBERSHIP DRIVES BEGIN

Froker advised Binversie, Lutzke and the other members of the fact-finding mission that the cooperative could not be formed until at least five hundred farmers representing ten thousand cows collectively signed a five-year contract. That multiyear commitment would ensure a steady milk supply. But first, they needed a name for the emerging group in order to hold an inaugural membership rally.

"In a brainstorming session one evening, a committee member, no doubt thinking big and of the large condensery in Manitowoc, A&P (Atlantic & Pacific), rather facetiously suggested 'Ocean to Ocean,'" wrote Torgerson in his book *Building Markets and People Cooperatively*. "Immediately, Wally Freis, committee chairman, blurted out 'Lake to Lake.' Thus, the name Lake to Lake Dairy Cooperative was chosen," continued Torgerson, who was called to the armed services in June 1945 but discharged from the navy before year's end, as Japan surrendered that September.

Manitowoc County and its nearby lake counties of Brown, Calumet, Door and Kewaunee would be the recruitment focus for the new dairy co-

"For Goodness Sake, Buy Lake to Lake Dairy Products" became a billboard campaign of sorts. However, this roadside campaign was featured on silos throughout northeast Wisconsin. This replica can be found on the silo at Heritage Farm in rural Kewaunee County. *Author's collection.*

op in Wisconsin's northeastern milkshed. The lake counties were bordered by Lake Winnebago on the west and Lake Michigan on the east. Years later, dairy farmers in Fond du Lac, Sheboygan and Winnebago Counties were invited to join the cooperative based in the milk-rich northeastern Wisconsin milkshed.

Nearly five hundred dairymen attended the first rally at Oukers Hall in Silver Lake. As a result of that first meeting, eighty-eight Manitowoc County dairymen signed five-year contracts on the spot. That night, three carloads of Door County dairymen also joined the discussion. The representatives, hailing from ten of the county's townships, were so convinced of the idea that they decided to hold a membership drive the very next night and invited Froker to speak at their event. Paul Wolski, a Door County extension specialist, helped organize the group in short order by making phone calls alongside the dairymen.

On the night of Door County's first membership drive, its county board adjourned their meeting early because they could not hold quorum, as interest was running so rampant as to the potential of a new milk processor in one of Wisconsin's most logistically challenging regions for shipping milk. The rally was held at Sturgeon Bay's Sawyer School.

Signing the contract was one matter. By placing one's John Hancock on the dotted line, each dairy farmer also made the financial commitment to purchase ten dollars in stock for every milking cow in the barn.

More meetings ensued. Eventually, the five-county steering committee convinced many more than the prescribed 500 dairy farmers to join, so the goal was expanded. The Manitowoc County steering committee set a new goal of 500 farmers alone, and Door County's leaders committed to 400. This took place in an era when USDA's Ag Census data recorded Wisconsin as having 154,000 farms with commercial milk sales from its 2.4 million dairy cows. In all, Manitowoc County had 3,400 dairy farms and Door County had 1,765, according to the 1945 Ag Census. By comparison, the entire state of Wisconsin had 6,500 dairy farms holding permits to sell milk in January 2022 with 1.3 million dairy cows.

To the awe of many, Door County crossed the finish line to reach its membership goal. Lawrence Johnson, a dairyman who went on to be the co-op's longest serving director, signed up the highest percentage of dairy farmers for Lake to Lake in Door County. The Claybanks township was Johnson's signup stronghold.

Manitowoc was close behind with the instigator himself, Henry Binversie, single-handedly signing up one hundred dairy farmers. To honor his hard work, Binversie's farm earned number 1—as in, all his farm's milk cans bore the number 1. In those days, each farm, and its milk cans, was assigned a patron number to identify the milk farm location.

Meanwhile, it was slower going in Brown, Calumet and Kewaunee Counties on supporters. This all took place even though there were no plant construction plans, no management in place and no definitive answer to the most important question, "When will farmers begin shipping Grade A milk?"

In 1946, the fledging cooperative and its members shipped 1.1 million pounds of milk. For perspective, my parents' sixty-cow herd outdid that total in its best year decades later. In 1946, this co-op was a newborn infant.

By 1947, Lake to Lake's charter members shipped 10.4 million pounds. That happened to be the year Elmer and Julia signed on the dotted line and pledged $10 per cow for a total of $400 for their forty head.

NOT SMOOTH SAILING

Despite the organizing committee's early success, there were stressful moments. "Opposition meetings were held to thwart the new 'menace,'" recalled Torgerson, with the menace being farmers taking control of their own destiny. "At School Hill, naysayers misrepresented the facts so badly that young farmer Orin Berge was moved to peel his coat and challenge outsiders." The challenge was presumably a fistfight. "At a meeting in Menchalville, Lake to Lake committeemen were shocked when called communists," Torgerson continued. "Feelings ran so high it even reflected relationships at the local church." Yes, fellow parishioners quit speaking to one another.

"Local business groups attempted to slow formation of the cooperative, too. Most noteworthy was the petition to the (Manitowoc) County Board of Supervisors by the Two Rivers Community Club," Torgerson recalled in his book. This opposition largely came from privately held cheese plant owners and their relatives. "The club charged that I was being overzealous in promoting the cooperative. They wanted me to refrain from such activity or be removed [from my county extension role]. Farmers defended me and the petition was buried at the County Board."

Torgy, as he was known by the farmers, was not a man who backed down from a fight. As a student at the University of Wisconsin, he boxed his way to the National Collegiate Athletic Association's (NCAA) Light Heavyweight Boxing Championship. He could literally throw a punch; however, the Torgy punches in this fight came in the form of moving the cooperative forward for the sake of farmers.

When it came time to hire the first manager, three seasoned candidates turned down the fledging Lake to Lake. That's when the five-county steering committee regrouped and discussed a candidate in their midst, long on passion for creating Lake to Lake though he was short on management experience. After a unanimous vote, Truman Torgerson became the cooperative's first and only general manager, serving from 1947 to 1980.

Torgerson insisted on a united vote from the group. That was the only way he would accept the position. "The group did, and to this day the ballots recorded on torn bits of paper mounted on a sheet of paper are in my possession," Torgerson wrote of his hiring.

Suddenly There's More Milk Money

An amazing thing happened once it became evident that Lake to Lake was going to happen. Even though the fledgling cooperative was $1,501 in debt after its first full year in operation, it managed to raise the area's milk prices. The county went from being $0.08 under the state average to $0.16 to $0.18 above prices paid in western Wisconsin in just one year. The competition from the new business kid on the block was good, and all farmers, whether cooperative members or not, benefitted.

The cooperative's cheese plant in Kiel opened in 1948. That's when the milk trucks began to roll toward the southern Manitowoc County cheese facility—27.1 million pounds of milk in 1948, 93.6 million pounds in 1949 and 155.3 million pounds by 1950.

That very next year, Lake to Lake completed an ultramodern receiving station to collect milk in Sturgeon Bay. It opened in January 1951, and jubilation sprang forth throughout Door County's northernmost reaches. Members north of Sturgeon Bay began to have their milked picked up on a daily basis—some farmers had been waiting five long years for this opportunity to have a consistent and reliable milk check. Remember, these were the days before Interstate 43 was on the drawing board or Highway 57 from Green Bay to Sturgeon Bay became a four-lane expressway. It was definitely a much slower roll in those days from Wisconsin's long peninsula to the co-op's processing plants in Denmark and Kiel.

That same year, the cooperative rolled out its first sales truck. "A spiffy white with red and blue Lake to Lake logos adorning the side panels,"

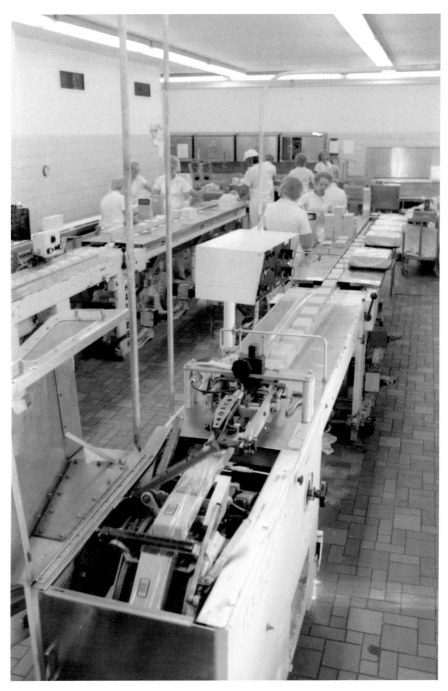

Lake to Lake's Kiel, Wisconsin cheese plant eventually became the first in the country to earn USDA's Grade AA Shield—the highest standard for cheese making. *James Baird*.

said Torgerson. "Roman Dennison was our first driver salesman," he continued, noting that the charismatic Dennison later parlayed his skills into being elected mayor of Green Bay. "We enjoyed success getting our butter and cheese into stores," wrote Torgerson of getting that first route truck on the road.

Torgerson also chose the milk haulers and signed contracts. He insisted that each hauler wear a white shirt and white pants to reflect the purity of the milk. He wanted to be a premier dairy processor.

"As a little girl, I always remember our milk hauler, Vic Grimm, Sr., wearing white clothes at each pickup," said Elmer's daughter Rosalie years later. "Vic would deliver cheese and butter whenever Mom placed an order. The cost would be deducted, at a patron's discount, off our milk check."

Torgerson noted that average purchases by members were 122.5 pounds of cheese and 50.4 pounds of butter each year. "Both were far above the national per capita consumption," he noted.

By the time Rosalie was able to recall these events, Vic had mounted a stainless-steel bulk tank on his truck's chassis. Her parents, Elmer and Julia, were more than willing to upgrade their farm's facilities to meet the Grade A milk shipment standards and capitalize on the additional premiums for their farm's milk. So, Elmer went about bringing his milk house up to the Grade A standard, and the milk cans became decorations. By the time Rosalie reached her eleventh birthday in 1960, Lake to Lake had converted all its Grade A herds to bulk pickup. This was a remarkable achievement in its day.

At the cooperative's tenth annual meeting, Rudy Froker, with a smile on his face, came back to speak to the group and laid out some lofty goals for the cooperative. By this time, the visionary economist who drafted the Lake to Lake idea had earned a new title—Dean R.K. Froker became just the sixth person to lead the College of Agriculture at the University of Wisconsin since Governor Hoard founded the college in 1889. Froker would serve in that role until 1964.

FARM SIGNS SPROUT UP

The co-op, rather proud of its roots, initiated a sign program in 1953. At that moment in time, Milwaukee's Golden Guernsey and Pennsylvania's Lehigh Valley Cooperative had been the only known farm sign programs in the United States. Thus, Torgerson began a new tradition throughout northeast Wisconsin's countryside.

Lake to Lake initiated its farm sign program in 1953, and soon all milk buyers followed suit. The red triangles on this sign indicate processing plants, while the sperm whale in Lake Michigan indicated early co-op members. *Author's collection.*

"So effective was this program that one by one all the other dairies adopted it. By the 1980s, nearly every dairy farm had a sign," noted Torgerson.

In 1961, the cooperative set a new milestone by becoming the first cheese manufacturer in the United States to qualify for USDA's Grade AA shield. Consistently producing the highest-quality cheese took Lake to Lake "from an art to a science in cheesemaking," said Torgerson. That same year, the Chicago Board of Health approved every-other-day milk pickup. From that moment on, milk could be picked up every other day, allowing a savings for all farmers and their cooperative when compared to the every-day pickup standard that had previously prevailed.

Throughout the 1960s and the 1970s, the co-op continued to grow. It became a premium cheese brand throughout California, especially in the Los Angeles market. Land O'Lakes, in control of the overarching marketing agreement, insisted that Lake to Lake focus on the West Coast. The timing was perfect for the northeast Wisconsin cooperative, as America's population, and many of its sports teams, were moving west.

By 1980, the 1,285 Lake to Lake members voted to merge into the larger Land O'Lakes Cooperative. Lake to Lake had prowess in cheese and Land O'Lakes was its equal in butter. That year, Lake to Lake achieved sales of $130 million. And by mere coincidence, that's the year Elmer and Julia sold their cows. They sent their last load of milk to the Kiel plant on October 30, 1980. It was Julia's sixty-second birthday. Like her mother, Anna, Julia milked her last cow on her birthday.

On January 1, 1981, Lake to Lake became a division of Land O'Lakes.

"It takes a giant to compete effectively with the giants in the food business," said Torgerson of the merger. "In these days of extremely high distribution costs, single line companies are at a disadvantage. To be effective in research and product development takes millions of dollars that no Wisconsin cooperative has. It is time for Lake to Lake to move ahead, to unify with Land O'Lakes so we may strengthen each other and avoid duplicating each other's efforts in the marketplace."

Lake to Lake continued to be its own division of Land O'Lakes. But by 1993, the Lake to Lake name had faded away, although its signature plant in Kiel, Wisconsin, continues to produce some of the world's best Cheddar under the Land O'Lakes brand name.

15

FOR THE LOVE OF ALLIS

Two new creations rolled off assembly lines in 1949, and both were anxiously anticipated arrivals at Elmer and Julia's farm that January. The first creation came from God's assembly line when a baby girl named Rosalie Mary Pritzl was born to the farm couple on January 17.

Within days of Rosalie's arrival, Elmer drove a brand-new Model WD Allis-Chalmers tractor onto the farm. That WD tractor rolled off the assembly line of Wisconsin's mightiest manufacturer: the Allis-Chalmers Corporation, headquartered in the greater Milwaukee manufacturing epicenter of West Allis, Wisconsin.

The city earned its "West Allis" name in 1902, just two years after the company's corporate leaders purchased a one-hundred-acre site in the little hamlet originally called Honey Creek, in Milwaukee County's western township of Greenfield. The company made the move because it had outgrown its first location, where it built its industrial complex that would eventually employ twenty thousand people during the height of World War II.

The "West" was presumably added to Allis since the new plant was four miles due west from the company's original plant located in Walker's Point near downtown Milwaukee. And the West denotation is noteworthy, as no other Wisconsin urban center bears the name Allis.

As that new plant was being built, the Allis-Chalmers Corporation was already growing into an industrial juggernaut. "By 1900, Allis engines were hoisting rock in South African diamond mines, reducing ore to iron in Austrian blast furnaces, running flour mills in China, and supplying the motive power for New York City's transit systems," wrote John Gurda in *The Making of Milwaukee*. "The New York engines, which peaked at 12,000 horsepower, were the largest steam engines ever built."

THE FARM'S DYNAMIC DUO

All told, Elmer and Julia's new duo introduced in 1949—Rosalie and the WD tractor—would go on to accrue 136 years of service to the farm by the time this book rolled off the presses. The nod for "Longest Years of Service Award" goes to that WD Allis-Chalmers tractor, which eventually earned the nickname "Grandpa's Tractor" in 1981. That's the year when Rosalie returned to the farm with her husband, Randy, and children, Corey and Angela, to become the fifth and sixth generations to run the farm.

That's also the year Grandpa Elmer "retired" and moved to town. Anyone who knows a farmer already knows farmers never really retire. Hence, Grandpa's tractor stayed on the farm to do Grandpa's work in farm fields and the woods.

As for Rosalie's "Service Award"? Rosalie and that tractor were separated by only twelve years of service, as she and Randy farmed in the nearby Brillion Township from 1969 to 1981.

Which was Elmer's favorite?

To be fair, if one asked Elmer what his favorite arrival was in January 1949, he surely would have said his baby girl, Rosalie. But there would be very few items ever to rank above that WD Allis-Chalmers—other than his children, wife and Christian faith.

The 1949 Allis-Chalmers WD became the instant workhorse on Elmer and Julia's farm. "It was the first rubber-wheeled tractor purchased by my father to run his 215-acre farm. After my husband and I purchased the then 141-year-old family farm, my father kept this specific tractor so he could still work on the farm with equipment he was comfortable operating," said Rosalie. "This was the tractor he most enjoyed. He used it to log trees for lumber, seed alfalfa, side-rake hay, haul wagons and an assortment of other jobs," continued Rosalie. "It has seen so much service that its engine has been overhauled three times and is on its second set of tire rims."

Those tire rims present a unique story. While rubber tires had been an option on tractors for well over a decade, all raw materials were rationed during World War II, with rubber being one of the most rationed

When the Allis-Chalmers WD tractor arrived at the Pritzl farm in 1949, it became an instant workhorse. Its arrival also signaled the transfer of work from animal horsepower to mechanical horsepower. *Author's collection.*

commodities. That caused many farmers to put off the purchase of a tractor with rubber tires late into the 1940s. If they were fortunate to have rubber tires during the war, a patch kit was a must, as there were no new tires to be had.

ELMER'S ORANGE FARM

Most farmers have their favorite color. There's John Deere's iconic green and yellow and the classic reds for Farmall, International Harvester and today's Case-IH.

When it came to Allis Chalmers, the color was orange. Persian Orange, to be exact. The color dates to 1929. That's when Harry Meritt, the manager of the Allis-Chalmers growing tractor line, took a trip to California. Having come along a brilliant field of blooming poppies—orange poppies, to be precise—Meritt believed he found the color to better market the tractor line. Reports indicate those wild poppies could be seen for miles. That's how Meritt wanted to market the Allis-Chalmers line of farm equipment.

Up until that point, the Allis-Chalmers brand had been using a rather dull dark green enamel as a base and an accent of red for the wheels. On returning from that 1929 California trip, Meritt asked experts at the Pittsburgh Plate Glass Company to duplicate the color he found in that poppy field, reported Charles Wendel in his book *The Allis-Chalmers Story.*

"The majestic orange was an instant attention getter," explained Wendel. By the time Elmer became a loyal Allis-Chalmers customer in 1949, everything was painted brilliant Persian Orange as it rolled off the company's tractor line.

In addition to the WD, Elmer eventually owned a Model B, a Model C and two Model WD-45s. The Model B tractor went into production in 1937, and by the time the final Model B rolled off the West Allis assembly line in 1957, 127,186 tractors had been manufactured.

The Model C had a shorter run. From 1940 to 1950, Allis-Chalmers, which helped give Milwaukee the moniker "Machine Shop of the World," would produce 84,030 Model Cs. The WD tractor went into production in 1948, and by the time 1953 came along, there were 160,384 renditions working fields across America. The more powerful WD-45 had 90,351 models manufactured from 1953 to 1957. Two of those resided in rural Reedsville.

In its day, the WD tractor that graces the cover of this book featured state-of-the-art technology. Its wet clutch between the power take-off (PTO) shaft and transmission provided continuous PTO power totally independent of ground travel. The main clutch did, however, disengage both drives simultaneously.

Having pioneered the use of hydraulics, the WD tractor also had this feature. Additionally, the WD tractor came with three wheelbase options—a single front tire, dual narrow front tires or an adjustable wide axle version. Grandpa's tractor had dual narrow front tires, and with its short wheelbase and overall compactness, that tractor was easy to handle and could quickly make a turn at the ends of fields. In the woods, it could slip around and navigate trees like none other.

"My dad could fly on that tractor," recalled Rosalie from her childhood days. A four-speed transmission, electric starter, lights, PTO shaft and a hydraulic control system rounded out the features. Just in case the starter didn't work, the WD still came with a crank—as in crank start the motor. The total cost of Grandpa's version was just shy of $2,000.

By 1953, the last year of production for the WD, Allis-Chalmers incorporated its newly developed Snap-Coupler system into the WD tractor. With the Snap-Coupler came an entire series of implements developed especially for the use of this system, highlighted by Wendel in his extremely detailed book about the famous Persian Orange tractors.

Allis-Chalmers and Elmer Pritzl were synonymous. He is shown here giving his granddaughter Samantha Kable (née Krueger) a ride on one of his two WD-45 Allis-Chalmers tractors. *Author's collection.*

While Grandpa's WD tractor didn't have the Snap-Coupler, its two newer WD-45 cousins that joined the farm a few years later had them. As its name would imply, the WD-45 featured more horsepower—45 horsepower to be exact. It also came with power steering that made the long days on the tractor just a bit easier on the arms of the driver turning the steering wheel.

WELCOMED ADDITIONS

Beginning in 1949, the Allis tractors began to take over the workload on Elmer and Julia's 215-acre farm. As the orange and black starting team of five, which matched the colors of nearby Reedsville High School's basketball players, came into form, those tractors would eventually plant crops, bale hay, haul wagons and do all the field work. That included plowing fields in open cab tractors on cold fall nights. Despite those brisk nights when Elmer was bundled up in his hunting clothes after milking the cows, he was happy as he no longer had to walk behind a "horse's ass."

"Can you tell me what was so special about walking behind a horse plowing a field all day?" Elmer shared with yours truly in a rhetorical question years later. "The happiest day in my farming career is when I bought the tractor and sold the horses. Tractors were far more predictable."

It was that frank discussion and rather realistic observation about getting work done that made Grandpa Elmer a contributing member of the Greatest Generation. For the most part, today's horses serve recreational purposes, not as beasts of burden. Hence, the last team of draft horses left the farm forever in the mid-1950s, and those empty stalls were repurposed for milking a few more cows.

In addition to the five Persian Orange tractors and a pair of matching moldboard plows, Elmer had a Model 60 All-Crop Harvester. The All-Crop Harvester was another feat of Allis-Chalmers engineering that eventually spelled the end for the grain binder and threshing machine.

That piece of equipment was purchased right after World War II, 1947 to be exact, and it greatly sped up combining grains and clover seed. "The combine was a blessing!" wrote Julia years later in her journal about this piece of equipment.

Prior to its purchase, Elmer, like all neighboring farmers, relied on the time-tested threshing crews. However, that process took loads of people power—as in extra people working on the farm and Julia and her mother, Anna, cooking all day to feed the workers breakfast, lunch and dinner until all the crops were harvested. "I still remember the threshing crews we had to feed, and they could really pack it away," Julia recalled, noting the *it* was food. "I was glad when Elmer bought a combine and that job was eliminated," with the job of course being cooking meals for the team of hungry men.

Once the All-Crop Harvester arrived on the farm, the harvest sped up. In flipping through the vintage owner's manual, one can still smell grease and oil waft from the pages. Elmer's handwritten notes also indicate alsike clover, page 26; oats, page 30; and barley, page 23—in that order—were the most important grain and seed crops on the farm. The clover and barley were cash crops, and the oats were stored in the granary and later fed to the farm animals.

It's Your Turn

To this day, farm life is a constant rite of passage. As children grow, often more is expected of them. One of those occasions took place in the early 1960s when the oldest in the family, Jacque, was underperforming when it came to cultivating field corn. Word has it Elmer, a farming perfectionist and a penny pincher, was hopping mad on the occasion, as his oldest daughter was "plowing up too much corn."

That's when Elmer decided to give young Rosalie a try at the task of surgically driving the Model C Allis-Chalmers, with its tractor-mounted cultivators, through the field of young corn plants. The cultivating process turned the ground between the corn rows to disturb or even kill the weeds without "plowing up the corn."

Back in those days, nearly every farm was "organic" by today's standard. Herbicides, pesticides and commercial fertilizers were just beginning to enter the scene. Cultivating was the practice to give corn the upper hand over the weeds.

After some instructions, Elmer turned Rosalie loose on the Model C that the family had dubbed the "Gopher Chaser." That's because by the 1960s, its only purpose was cultivating corn and chasing gophers, since it lacked the horsepower to do most other farm work.

Elmer rolled up a cigarette and watched young Rosalie cultivate corn for a few rounds. Convinced his corn was living and the weeds were about to die, he headed for home, leaving Rosalie with the rather mundane task of driving back and forth methodically across the entire cornfield.

When lunch rolled around, Rosalie was nowhere to be found. Elmer sat down at the lunch table with Julia and children Jacque, Butch and Annie.

In addition to cultivating corn, Rosalie Geiger (née Pritzl) helped her father combine small grains and clover. *Author's collection.*

Julia was not too pleased, as lunchtime was lunchtime. Everyone showed up at the same time, on time.

While Julia ruled the house on nearly every day of the year, on that summer day Elmer clearly got his way. He hadn't given Rosalie a watch to indicate when she should come home for lunch. That was done on purpose, as Elmer clearly wanted that corn cultivated before it was too tall to allow the tractor to pass over the young field of maize. And based on Rosalie's rookie performance, it became her job until her wedding day in 1969.

INTO THE WIND

Even though there was not a farm dispersal when Elmer and Julia transitioned into retirement, some of the orange tractors had to go when Randy and Rosalie arrived on the farm in 1981 with their own fleet, as the couple had been farming for twelve years on their own. That's when the Model B and Model C were sold privately. One of the WD-45 tractors was sold to Randy's brother, Albert Geiger, who also purchased the couple's Brillion-area farm.

The second WD-45, with a trip bucket loader, stayed on the farm for about five more years until the Geigers installed one of the very first manure basins in northeast Wisconsin. That's when the trip bucket loader—which required the operator to pull a lever, causing a cable to release the bolt and

In July 2008, Elmer Pritzl's WD Allis-Chalmers tractor served as the sentinel when family and friends arrived to pay their final respects to a man who worked the land all his life. *Author's collection.*

Bruce Stetson began working for his father, Arlin Stetson, at the family's Allis-Chalmers dealership early in life. Bruce, who later took over the business, is holding an aerial photo of the business circa 1968. *Author's collection.*

allowing the bucket to flop open by gravity to dump its load—became a limiting factor to cleaning out the manure pit. Like the horses in the 1950s, that's the year the WD-45 hit the highway, too. Call it progress in a move to get more work done using less time.

However, Grandpa's WD Allis-Chalmers made the final cut and has continued to be a permanent member of the farm team, filling some rather unique roles. It's also been the farm's sentinel in many ways, even serving as a greeter at Elmer's wake and funeral.

"My father passed away on July 14, 2008, just a month shy of his ninety-first birthday," recalled Rosalie. "In honor of his life, we parked his pride and joy outside the funeral home and at the graveside as my dad's family and friends always remembered him on this tractor and on the other four Allis-Chalmers he owned," explained Rosalie. "Even the parish priest and sister thought it was a fitting honor for a man who worked the land all his life."

Grandpa's tractor is now on its third life, having received a full overhaul by Bruce Stetson, whose family had been a long-time Allis-Chalmers dealer in nearby Menchalville, Wisconsin. The Stetsons had that dealership until the unfathomable death of the mighty Allis-Chalmers Corporation.

16

FOR THE DEATH OF ALLIS

In its day, Allis-Chalmers was the brightest star in Wisconsin's "Milky Way" of manufacturing. No star shined brighter than the home of the Persian Orange. However, that massive red supergiant star, the largest Wisconsin had ever seen, eventually burst into a supernova, never to shine again.

Formed in May 1901, "A-C" eventually employed more people than any other Wisconsin enterprise. By the 1920s, the company offered two weeks paid vacation for employees, and its leaders even built a company hospital to lure workers to its manufacturing center.

As a result, the company rather quickly morphed into a manufacturing magnet, pulling both employees and employers to its global hub in West Allis. Consider that, in 1910, just eight years after West Allis incorporated as a city, it had a population of 6,600 people. By the start of 1930, the population had grown 422 percent to reach nearly 35,000 people.

Allis-Chalmers, and its deep talent pool of people, focused a great deal of energy on research and development, so much so that the company was highly respected by the U.S. War Department, the predecessor to the U.S. Department of Defense. When World War I rolled around, Allis-Chalmers was manufacturing marine engines, naval artillery and shrapnel shells for Uncle Sam.

As that war ended, Allis-Chalmers continued with its hard-charging growth trajectory. When combined with the output from the nearby International Harvester, the A-C and IH tandem, even though competitors, turned out more tractors than any other city in the world by 1929.

The 1931 purchase of the Advance-Rumely line proved pivotal to Allis-Chalmers. Not only was the Rumely line internationally known, but it was also highly respected by many farmers and machine men. In addition to that respect for expertise, the Advance-Rumely line also came with a vast dealer network. It's at that moment that Allis-Chalmers finally had the footing to compete with International Harvester, John Deere and J.I. Case. And compete it did.

Transformed Tires

The very next year, Allis-Chalmers engineers worked with a somewhat reluctant Firestone Company to mount air-inflated pneumatic tires onto a Model U tractor belonging to Albert Schroeder of Waukesha, Wisconsin. In April 1932, the closed-to-the-public test was successful. In fact, it was so successful that the world's first public demonstration of a tractor with air-filled tires took place on Labor Day.

By October 13 of that same year, Allis-Chalmers announced it would offer inflatable rubber tires as standard equipment because those tires resulted in a one-third improvement in fuel economy, and that led to getting one-quarter more work done when compared to tractors rolling on steel tires with lugs. As a testament to this remarkable achievement in engineering, the original Model U Allis-Chalmers tractor bearing those inflatable tires stands on display in the Wisconsin Historical Society's Stonefield Historic Site in Cassville, Wisconsin.

Despite their great success, those inflatable tires, which replaced the steel wheels, did cause a safety issue, as tractors now had a higher center of gravity, which caused them to roll over a bit easier. To remedy that matter, Allis-Chalmers engineers went back to the drawing board and ultimately applied for a patent with the U.S. Patent Office on March 20, 1937. Engineers filled a hydromatic tire with fluid to greatly improve safety, as the ballast lowered the center of gravity.

By the next year, all Allis-Chalmers tractors had fluid-filled tires. To quote an A-C catalog of the era, "Instead of being filled entirely with air, as in the past, these tires are three-fourths filled with a special freeze-resistant liquid ballast that has a preserving effect upon the inner tube tire. In addition to providing the extra weight necessary for efficient traction, the hydromatic tires also lower the center of gravity for improved safety." Eventually, every tractor company followed the lead of Allis-Chalmers.

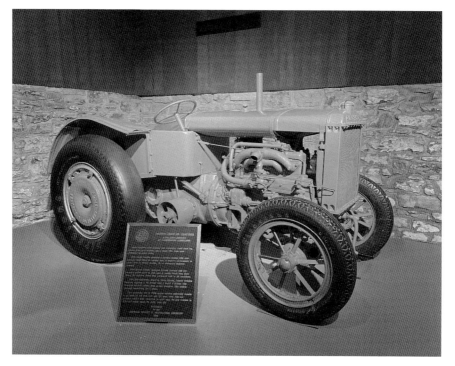

On display at the Stonefield Historic Site in Cassville, Wisconsin, this 1932 Model U Allis-Chalmers was the first in the world to be fitted with inflatable tires. By 1940, 95 percent of all tractors were sold with rubber tires. *Author's collection.*

Back to the legendary achievements of Allis-Chalmers. During the 1930s, in the depths of the Great Depression, A-C crafted the four generators, each capable of turning out 115,000 horsepower, for the famed Hoover Dam. That dam was the world's largest dam in the first decade of operation and boasted the world's largest hydroelectric plant. That government contract led to an even larger role during World War II.

SOLDIERS STAND ON GUARD

Allis-Chalmers became one of the nation's most critical strategic military contractors. That's because A-C produced more nuclear equipment, by pound, for the Manhattan Project than any other American company. So important was the Allis-Chalmers know-how to the Allied war effort, President Franklin Roosevelt made a surprise in-person visit to the Allis-Chalmers plant on September 19, 1942. The appearance, censored due to

After making stops at Chrysler and Ford plants in nearby Michigan on September 18, 1942, President Franklin Roosevelt visited the U.S. Naval Training Station near Chicago and the Allis-Chalmers Corporation on September 19. FDR would travel nearly nine thousand miles on his inspection tour from September 18 to October 1, 1942. *Wisconsin Historical Society.*

its importance to national security, was reported a month later by the *New York Times.*

"Uniformed soldiers stood guard at the entrances of most defense plants, and employees were generally denied access to all but their own work areas," reported John Gurda in his book *The Making of Milwaukee.* "Factory workers at the Allis-Chalmers plant were instructed not to talk to each other."

Just three years after FDR's visit to West Allis, pilot Paul Tibbets and his *Enola Gay* dropped Little Boy on Hiroshima on August 6, 1945, and three days later, pilot Charles Sweeney and his *Bockscar* dropped Fat Man on Nagasaki. Allis-Chalmers know-how was at the heart of it all as Japan waved the white flag to end the war.

In the 1950s, 56 percent of the greater Milwaukee area's workforce was engaged in manufacturing, one of the highest concentrations in the United States. Allis-Chalmers topped the leader board in employing the most people.

President Franklin Roosevelt is looking through the rear window while Wisconsin governor Julius Heil applauds. Walter Geist, Allis-Chalmers president, is seated near FDR, with James White, Allis-Chalmers works manager, in the front seat. Grace Tully, FDR's personal secretary, recorded, "This is the first place we had hit where, to the employees, the President's visit was a cold, complete surprise." *Wisconsin Historical Society*.

By the time 1961 came around, A-C had created an X-Ray machine to treat cancer. Known as a cyclotron, that first-of-its-kind instrument went to work at the Washington University Medical Center. The mighty Allis Chalmers Corporation continued to chug along in the 1960s and the 1970s.

FROM STAR TO SUPERNOVA

As a new decade started, Allis-Chalmers still laid claim to being Wisconsin's brightest manufacturing star. Sales peaked at $2 billion in 1980, and profits approached $50 million. However, that was the last banner year Allis-Chalmers would spend in the black. That's also the year Wisconsin's supernova began to form in West Allis. A supernova is the biggest explosion that humans have ever seen. Its takes place when a star explodes into billions of pieces. The supernova itself is the "last hurrah" or "final gasp" of a massive dying star. And make no doubt, Allis-Chalmers was the most massive manufacturing star Wisconsin had ever seen. In the early 1980s, it began to belch out its final gasps.

In 1981, Allis-Chalmers lost $29 million. The next year, it lost a state record $207 million. By 1984, it topped that number by hemorrhaging $261 million. In fact, red bled out everywhere from West Allis, as it lost over $500 million in a four-year window. Share prices tumbled from 1979's $38 to a staggering $3 by 1985.

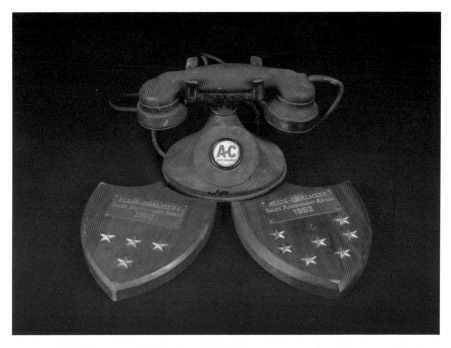

When Allis-Chalmers gasped its last breath, shockwaves went out into the countryside. At the Stetson Allis-Chalmers dealership in Menchalville, Wisconsin, the phone no longer rang for new equipment orders. *C. Todd Garrett.*

On June 29, 1987, Allis-Chalmers burst into a supernova as it did the unthinkable—filed for bankruptcy. The epic failure hit the Milwaukee market hard, as 1980 began with the region having eleven Fortune 500 companies. By the end of the decade, that number had shrunk to six. In addition to the demise of the "Great Persian Orange Star," other headliners were Pabst and Schlitz—a pair of legendary brewers.

While beer may have made Milwaukee famous, it was its metal trades that employed the masses who bought that beer. At their zenith, Schlitz, Pabst, Miller and other brewers employed 7,100 people. That was just 2 percent of Milwaukee's workforce. Meanwhile, the Machine Shop of the World lost 56,000 manufacturing jobs from 1979 to 1983. That was a quarter of the total employment in the greater Milwaukee area, as manufacturing jobs fell from 220,000 at the close of the 1970s to 164,000 by 1983. Allis-Chalmers hemorrhaged the most jobs.

WHY THE EXPLOSION AND EXTINCTION?

In its heyday, Allis-Chalmers ruled the manufacturing landscape. In 1904, the company adopted a new slogan, "Ours the Four Powers: Steam, Gas, Water, and Electricity," after it purchased Bullock Electric Manufacturing Company. In those days, the company was America's largest manufacturer of steam engines. While the company's largest market share was in steam engines, the four-pronged approach also helped it become one of the world's largest heavy equipment manufacturers, too.

In its heyday, Otto Falk, the longtime president of Allis-Chalmers, called his company "a sleeping giant." While a giant due to its collective business volume, it was no longer a giant in any one production category after steam engines went the way of the dodo bird.

In fact, the four-pronged approach eventually led to its demise because the Allis-Chalmers Corporation did not hold a market lead in any category once steam engines lost favor to the internal combustion engine. John Deere and International Harvester sold more tractors, General Electric and Westinghouse had a far bigger market share in electrical machinery and Caterpillar, a smaller company in the 1950s, had become the dominant force in heavy machinery by the 1980s. There were no first-place blue ribbons in sales for team orange. And with those second-, third- and even fourth-place rankings in categories, higher distribution and marketing costs and lower productivity began eroding the company's financial position.

"The company relied on volume and diversity for its success, a strategy that worked for generations," observed Gurda. As the 1980s arrived, inflation, high interest rates, a struggling agricultural economy and other devastating economic forces converged to create the West Allis supernova. And the Allis-Chalmers bright star went forever dim on June 29, 1987.

Like an exploding supernova, pieces of its star still exist in the universe if one looks carefully. You still spot some Allis-Chalmers tractors working on farms, on parade at tractor shows and even a few at tractor pulls. The Hoover Dam still turns out electricity from its Allis Chalmers generator base. And the Persian Orange terrace chairs at the University of Wisconsin's Memorial Union still recall an area when Allis-Chalmers Persian Orange was the brightest color in all the Great Badger State. And if you drive past Elmer Pritzl's farm, you can still, on occasion, see his grandson and his WD Allis-Chalmers working on the farm.

17

HE MAY HAVE BROKEN FEDERAL LAW

One interesting mystery on the family homestead is woven into the farm's limestone rock wall. The wall runs parallel to a portion of the farm's quarter-mile-long driveway, and it shields the farmhouse from vehicle traffic—as if the home's twelve-inch-thick cement block walls need any further protection. A small army could use that home as a command post and be shielded against small arms and artillery fire.

Nestled among the interlocked limestone rocks—ledge rock and stones that number in the thousands—is a copper disk embedded into aging concrete. That disk, and its mortared base, have no business being in that wall.

Construction on the limestone wall started in the 1920s by Julia's father, Red John Burich. Years later, after Red John's death, his son-in-law Elmer Pritzl completed its construction. Field stones and limestone pieces from the area's Niagara Escarpment were picked one by one and placed either in wheelbarrows or on hay wagon racks as those rocks "grew up" from the farm fields each year. Then the stones were assembled into a freestanding wall.

No Business Being There

In searching for an answer to the copper disk mystery, a few lines from one of my favorite movies leaped into my mind. Instead of the movie's mysterious

Elmer Pritzl continued building the limestone stone wall adjacent to the farmhouse after his father-in-law, Red John Burich, passed away. *Author's collection.*

black volcanic glass being situated in a long rock wall under an oak tree, the copper disk was located along a long rock wall under an ash tree. One of the movie's leading characters involved another man named "Red."

"It's got a long rock wall with a big oak tree at the north end. It's like something out of a Robert Frost poem," says the wrongly convicted prison inmate Andy Dufresne to his friend "Red," played by Morgan Freeman in the movie *The Shawshank Redemption.*

"Promise me, Red, if you ever get out...find that spot. At the base of that wall, you'll find a rock that has no earthly business in a Maine hayfield. Piece of black, volcanic glass. There's something buried under it I want you to have."

Well, there was nothing buried under this copper disk and its concrete base in this farm story. Upon further investigation, however, that copper disk had been in a hay field just as in *The Shawshank Redemption.*

What hayfield?

That remained a mystery.

WHY DID ELMER MOVE IT?

Upon closer examination, the copper disk and the concrete around it had been attached to a much larger piece of concrete. The edges were rough as if chipped away with a pickaxe or bludgeoned with a bulldozer blade.

While I heard many stories from Elmer and Julia over the years, this topic was rather void of conversation. "Can you believe that federal workers poured concrete in my hay field and set that marker there? The nerve of those guys!" exclaimed Elmer in a short discussion in the 1980s. He wouldn't divulge much more even after an inquisitive grandson peppered his grandfather with questions.

Elmer never did give up its original location and only quasi-confessed to removing the marker a short time after the concrete was cast.

While Elmer might have cussed its very existence as a farmer who toiled for decades to clear rocks and stumps from his farm fields, future surveyors likely cussed out Elmer, too. As it turns out, that copper disk was a survey benchmark. These geodetic marks were placed to indicate key points on the earth's surface. Elmer's farm field just happened to be one such point.

"U.S. Geological Survey Benchmark," read the words on the outer rim of the copper disk. "For information write the Director Washington D.C.," reads the next interior line. Dead in the center it reads, "Elevation 870 feet above sea level." Finally, it has a unique indicator, "TT 25 RGS," hand-stamped into it along with a date, "1951."

This U.S. Geological Survey Benchmark rests a few football fields from its original location. Back in the day, some of these disks were set into concrete and placed in farmers' fields. *Author's collection.*

From Benchmark TT 25 RGS, surveyors could complete land surveys for miles around. But first they had to find the benchmark. This one was long gone—smuggled into a limestone rock wall for safekeeping—just in case federal officials stopped by one day and Elmer needed to clear his name.

As it turns out, the U.S. Geological Survey generally set these marks for keeps. The federal agency often embedded cast metal disks on top of concrete pillars and set them into the ground. These benchmarks and the concrete that surrounded them were intended to be permanent, and disturbing them is generally prohibited by federal and state law.

Giving Up Its Original Location

The National Oceanic and Atmospheric Administration (NOAA) maintains a permanent record of all such benchmarks. After completing an online search, only five cousins to our mysterious benchmark exist in Manitowoc County. Those five benchmarks also were created in 1951 by the U.S. Geological Survey.

Why did so many disappear?

It's likely that the 1951 crew didn't use common sense in setting those benchmarks and other farmers pulled them up, too. No farmer in their right mind wants to hit a rock with farm equipment. That's why every spring comes with an annual rite of passage for area teenagers—picking rocks from farm fields.

To Elmer, that concrete marker was just another rock lurking in his field, waiting for its moment to damage his field equipment. The mystery was beginning to be solved.

Map Discloses Location

Postmark: *Silver Spring, MD. November 17, 1957.* A 12-inch by 9½-inch business size envelope was mailed to Elmer J. Pritzl, and it gave up the mystery to Benchmark TT 25 RGS. In the fall of 1957, it cost just four cents to mail its contents.

Now, Elmer J. and Elmer W. were two different individuals. Elmer John Pritzl was named after his mother Julia Pritzl's most-loved men: her father, Red John Burich, and her husband, Elmer Pritzl.

The fourteen-year-old Elmer John purchased two maps from the U.S. Department of Interior: one for the Chilton Quadrangle and one for the Reedsville Quadrangle. For map hounds and history aficionados, these maps may be among the most amazing treasures.

Each map details an area approximately ten by fourteen miles in area. Every major feature can be found on those maps created in 1954: Kasson's Holy Trinity Church, Kasson's St. Bartholomeus Lutheran Church, Rockland's Immanuel Church and a host of rural one-room schools such as Maple Grove, Carson, Alcott, Hawthorne, Long Lake and Marquette. The map even lists the abandoned rural schools. All these buildings have faded into history—just like Benchmark TT 25 RGS.

A closer look at Elmer Pritzl's farm on the same map also lists BM 870. Sure enough, "BM" is the Geological Survey's abbreviation for Benchmark and the "870" is the elevation indictor. U.S. Geological Survey TT 25 RGS is on the 1954 map and reconfirmed by its elevation of 870 feet.

Its location on the map?

Not in the nearby ditch. Not in the nearby intersection of Manitowoc or Hickory Hills Road. But right smack in the middle of Elmer's hay field that he worked years earlier to clear.

When that concrete showed up in the field one year, Elmer just went back to work clearing it as if it were any other rock that could maim his farm equipment.

The Modern-Day Benchmark

The farm's location and topography have made it a location for unique activities over the years. Lime kiln operators purchased land, blasted and tested its limestone. World War II glider pilots soared over its airspace during night training missions for an eventual D-Day landing. Flights arriving in nearby Appleton and Green Bay airports regularly come into view just after takeoff and prior to landing. The farmstead also was even a proposed route for Interstate 43. That never happened, as Lake Michigan neighbors successfully lobbied for the interstate near their cities.

The U.S. Geological Survey liked the farm so much that it came back forty years later to drop another benchmark on the family farm. This one sits in concrete on the top of Goodwin Road near the peak of "The Rock" overlooking Reedsville and the Collins Marsh. Its position is just shy of one thousand feet above sea level.

This time around, Elmer's grandson, the author of this book, who owns the land nearby, pledges to leave the DF 6009 U.S. Geologic Benchmark sitting in its position guarded by three orange posts.

18

IF I DIE AS I DINE

"**O** my God, I am heartily sorry for having offended Thee, and I detest all my sins, because I dread the loss of heaven, and the pains of hell; but most of all because they offend Thee, my God, who are all good and deserving of all my love. I firmly resolve, the help of Thy grace, to confess my sins, to do penance, and to amend my life. Amen."

That was quickly followed with, "Bless us, oh Lord, and these Thy gifts which we are about to receive from Thy bounty, through Christ, Our Lord. Amen."

Not your usual mealtime prayer, even at a devout Catholic family's home. The latter prayer is known to many as a traditional Christian mealtime blessing. The first prayer is the lesser-known Act of Contrition.

Those two prayers were coupled together at the Pritzl dinner table when Uncle Quiren and Aunt Beatrice came over with their brood in summer and late fall. That's because mushroom season had commenced, and Uncle Quiren, and anyone he could wrangle to go with him, had just spent hours foraging in Elmer and Julia's woodlots for mushrooms before dining at the dinner table. Quiren learned the art of mushrooming from his Bohemian relatives, and when he married Beatrice, Julia's sister, those woodlots also became his happy hunting grounds for his favorite fungi.

But why start a meal with an Act of Contrition?

If someone is not not an expert forager, one must be careful, as many wild mushroom varieties are mildly poisonous, and a few are downright deadly and can send one directly to an early grave.

When Quiren Sleger cooked up his mushroom bounty, Elmer and Julia insisted that their children express an Act of Contrition just in case anyone consumed deadly mushrooms. Shown here are (*left to right*): Rosalie, Elmer, Elmer John (*standing*), Annie, Jacque and Julia Pritzl. *Author's collection.*

The bottom line is, if in doubt, don't eat it.

As far as Elmer and Julia were concerned, neither must have thought Quiren had earned his "expert" field credentials on the mushroom front. Hence, they insisted on the Act of Contrition just in case anyone ate a deadly mushroom and died as they dined. The Pritzl parents wanted their children to have a clean conscience if they were about to meet Jesus.

That Act of Contrition apparently spooked the following generations to swear off wild mushrooms altogether. As a grandchild of Elmer and Julia, I clearly got the vibe they were not going to teach me about mushrooming even though Grandpa Elmer taught me to identify every tree species in the woods during winter by looking at bark and branches—without the aid of leaves.

OUR FIRST FORAGE

As I got older, my curiosity got the best of me. It only grew after my friend Greg cooked up a batch of morels one spring. After eating the tenderloin of mushrooms, I was at least hooked on this one mushroom.

The morel has a unique look. To this dairy science graduate, it looks like cow stomach number 2—the reticulum. That's the second of four stomachs and resembles a honeycomb.

In the spring of 2013, the conditions must have been just right for these mushrooms to grow, because my wife, Krista, called with a great deal of excitement in her voice. "There are morels growing under all the pine trees on our lawn!"

"What?"

"Yes, morels. I'll send you a photo."

Sure enough, they were everywhere. We picked several gallon pails full during the ensuing days. Then my co-workers came over for a cookout. Every one of them acted like Grandpa Elmer and Julia—no matter what I said to convince them otherwise, they were skeptical of the mushrooms' safety.

As I sauteed them in butter on the grill, Krista and I began oohing and aahing as we enjoyed the tasty woodland treats. One co-worker came over and asked, "Does your stomach hurt?"

"Nope. I saved these to share with all my co-workers. However, if you are not going to eat them, Krista and I will polish that entire skillet off."

And so, the first co-worker ate one. Then another. And within five minutes, the morels were disappearing like flapjacks at a northern Wisconsin lumber camp.

AN AWAKENING

This Act of Contrition story lay dormant in my mother's subconscious for decades. She must have gotten a bit nervous about my foraging desires and only shared the entirety of that story with me in 2021 as I amped up my interest in fungi. My interest grew after Krista and I bought a property in Wisconsin's Door County.

As fate would have it, our first neighbors were Matt Chambas and his partner, Jamie Mead. A delightful couple, this pair of foodies thoroughly enjoyed searching the woods for wild food and natural-growing plants to use as seasonings.

Matt Chambas, chef and forager extraordinaire, is shown with the fruits of his fall foraging for mushrooms in Door County. *Jamie Mead.*

After eating some of Matt and Jamie's mushroom bounty picked in the fall of 2021, "Corey the Kid" went into full force and started asking all types of questions about Wisconsin's cornucopia of mushrooms.

Matt, who is an expert and even serves his mushrooms via their Door County Underground dining experience, shared that to find mushrooms, you must have some great woodlots and be able to identify trees. Well, I have the tree part down in spades. One day, I hope to pluck mushrooms like the owners of Door County Underground—Matt and Jamie—and even invite them back to Quiren's happy shrooming grounds. One thing is for sure: if Matt is heading the foraging team, I'll take a pass on that Act of Contrition before dinnertime.

It's a National Sport

That Act of Contrition never caused Quiren to flinch. He just kept on picking his tasty mushrooms and lived to a ripe old age of eighty-three, while wife Beatrice lived to be ninety-five. Even though he was born in America, Quiren had learned the foraging arts from his Bohemian parents and grandparents. To learn mushrooming was a rite of passage in old-guard Bohemian families, as the Czech people have been picking mushrooms since the Middle Ages. In old Bohemia, mushrooms were the "meat of the poor," and the delicacies would be featured during Christmas dinners and New Year celebrations.

Mushroom picking is considered a national sport in the Czech Republic. So important are mushrooms to the modern Czech people that most forests remain freely accessible for mushroom picking.

It has been reported that two-thirds of the country's ten million people go mushrooming at least once a year. The average Czech household picks more than eighteen pounds of mushrooms each season. The National Czech and Slovak Museum & Library in Cedar Rapids, Iowa, even has murals of mushrooms throughout its campus area. There's also a St. Wenceslaus Catholic Church a little over a block away.

It doesn't get more Bohemian than "Good St. King Wenceslaus" and mushrooms. That further documents the importance of what mushrooming means to the Czech people. And as with many sports, it's also highly competitive.

You Can Tell a Lie!

Just how competitive is mushrooming? The following short story from a fellow Bohemian family native to the Reedsville area explains it the best.

"My mother was a good and honest person who could not tell a lie, except about picking mushrooms," wrote Bernard Kubale in his book *The Place to Meet Your Friends.* "Every fall, after the first frost, she would announce it was time to 'see if there are any mushrooms.'"

Children were welcome to join in her search with two conditions. "We could not tell anyone if we found them unless she gave permission," recalled Bernard. "In no event could we tell where we found them," added Bernard of his mother, Josie, whose family touches our family tree.

Now Josie, born Josephine Novak, was a spirited person when it came to mushrooms. Food was at times scarce, as her mother, Catherine, gave birth to twenty-one children and Josie was the eighteenth child and among four sets of the family's twins. If one of Josie's siblings found out about a mushroom location, word would spread like leaves falling on a forest floor in fall. Everyone would be picking Josie's mushrooms.

However, Josie's inquisitive children hadn't yet connected the dots and the reason for supreme secrecy. As a result, Josie's brood just asked more questions on mushrooming in the 1940s.

"What should we say if someone asks us if we found any mushrooms?"

"It depends on who asks," said Reedsville's Queen of Mushrooming. "If it's someone who doesn't pick mushrooms, you can tell the truth. If it's someone who picks mushrooms, like Sister Emma, just tell them we didn't find any."

That resulted in a Gatling gun of questions from the Kubale children.

"What if they know we found some?" the children asked.

"Then tell them we did, but don't tell them where we found them."

"What if they ask, what do we say?"

"Tell them you don't know," quipped back their mom.

"But if we know, that would be a lie. Isn't that a sin?"

"I suppose, but it's a venial sin," Josie responded.

"What if Father Koutnik asks? What will you say if he asks you?" asked young Bernard Kubale, who was elevating the stakes by invoking the name of the parish priest.

"I'll tell him we didn't find any," responded Josie.

"What if he knows you found some and asks where you found them?" continued the Kubale children.

"I'll tell him someplace other than where we went," said the normally upstanding Josie.

"But then you're lying to a priest," the children pointed out.

"I'll confess it the next time I go to confession," she stated.

"You'll confess to him that you lied to him?"

"He won't ask who I lied to. Now that's enough. Just do as I say," Josie said.

And so, the Kubale children did. However, it was to no avail. Everyone in the Reedsville community knew Josie Kubale was a mushroom picking all-star. Josie would even misdirect people by going the opposite way out of town. However, local mushroomers caught wind of Josie's misdirection. Knowing that, sometimes Josie just went straight to her favorite sources, which included the Mahnke, Busse and Novak woodlots.

Josie's conscience did eventually shine through, as Bernard reports Josie would give Father Koutnik a quart or two of canned mushrooms for Christmas.

Mushrooming is a rather bare bones sport. Outside of a good woodlot, one needs only a sharp eye, a knife and a container to carry your haul. The knife is so important, because every good mushroomer knows never to pull up a shroom. You may disturb the mycelium, and that would eliminate future crops.

The most expert mushroom foragers carry a mushroom knife, as its brush is useful for gently cleaning soil from the gills and stems. Its curved blade allows the forager to carve off any debris and soil. And if you are hunting the mushroom known as bracket fungus, the mushroom knife is just sharp enough to cut the fungus from the tree bark without causing damage. Also, in the old country, many Czechs carry wicker baskets that allow the mushrooms to breathe even after they are picked.

In Wisconsin, some of the most treasured mushrooms are the chicken of the woods, the hen of the woods, giant puffballs and, of course, the highly coveted morels. That's if you can find them. Like all sports, if you're a rookie or on junior varsity, start with a good coach, or you may have wished you prayed that Act of Contrition.

19

LET'S GO FISHING

Minnesota may boast of being the land of ten thousand lakes, but like it's four-trophy lead on Lombardi trophies for Super Bowl Champions, Wisconsin also has Minnesota beat on the number of lakes—over fifteen thousand. Given all those lakes, many Wisconsin residents like to go fishing and eat those fish, too.

Elmer loved his chosen vocation as a farmer. Because Elmer farmed on Hickory Hills Road, in Manitowoc County's Rockland Township, he also got hooked on fishing, too. While it's true that the Rockland Township derives its name from the numerous outcroppings of coral limestone of the Kettle Range, within this glacial moraine are nestled five beautiful lakes.

Those nearby lakes—Long, Becker, Bullhead, Round and Boot, in that order as far as Elmer was concerned—all offered great fishing options less than six miles from his farm. Plus, Elmer was drawn to fishing in part because he watched all his neighbors drive by and head to those fishing holes Monday to Saturday as he worked on his farm from his vantage point on his fields and the barn.

Being a dedicated farmer, Elmer rarely fished on any day other than Sunday. Often, he would go out on the lake after morning milking and church service. In the early days, he rented a wooden rowboat from the Kanter family who farmed near Long Lake, the county's second-largest inland lake. Those were the days prior to the four-and-a-half-acre public access developed by the Long Lake Advancement Association in 1963. Due to the era's limited boat traffic and development, the lake was much clearer then.

It's about that time that a very young Rosalie, apparently bored with the entire fishing situation because she was too young to hold a fishing pole, took off her brand-new straw hat and tossed it in the lake when her mom and dad were not looking at her. "I remember it was such a sunny, warm spring day. I watched as the hat sank all the way to the bottom of the very clear lake. That caused big problems for me!" she said as it related to the follow-up discipline from her parents.

About a decade after Rosalie's straw hat floated to the bottom of Long Lake, Elmer purchased an Alumacraft boat. Once the three oldest children graduated high school, Annie, the youngest, would accompany Elmer fishing, although she found it more fun to catch the sun's rays than fish from the lake. However, Annie continued to go, as Julia didn't want Elmer on the lake by himself.

During his farming days, Elmer was an old-school fisherman. He launched his boat into the lake from its trailer attached to the family's car, a 1958 Ford Fairlane 500, and in later years the farm's two trucks: a spring green 1952 Chevy and, in later years, a bright red 1970 Ford. Neither of those trucks had a radio, as that was too extravagant and an unnecessary cost. In the same vein, Elmer used wooden oars to traverse the lakes. Only deep into his retirement years would he buy a boat motor.

To offset the cost of that modern convenience, Elmer started a worm farm and no longer purchased bait. A special corn mash recipe was his worms' food of choice. Living in the village by that time, he got water from

This 1958 Ford Fairlane 500 would be the first vehicle to pull Elmer's fishing boat to nearby inland lakes. Elmer is flanked by daughters Annie and Rosalie. That Ford Fairlane 500 also had a police interceptor motor, and the Pritzls' eldest daughter, Jacque, could make that car "fly." *Author's collection.*

the farm, since the village's water was treated with chlorine and too harsh for his worms.

Rounding out Elmer's fishing equipment was a full line of bamboo poles—the best money could buy. When Elmer caught a big crappie, bluegill or perch, those poles would bend as if Elmer and his guests were deep-sea fishermen. Even a smaller sunfish could spur those bamboo poles into action.

Like Elmer, Julia also liked fishing and accompanied him on many fishing trips once the cows were gone. Children and grandchildren would go along, too, as would friends and neighbors, including Dan Dvorachek, who worked on the farm for Randy and Rosalie in later years. If it wasn't for Dan, Elmer might not have been able to go in his late eighties.

The love of fishing ultimately became one of the reasons the couple's son, Elmer John, known to nearly everyone as Butch, took a pass on running the family farm. Butch didn't want to watch boats go by on trailers while working on the farm. Butch wanted to go fishing more often than just on Sundays.

In the days before there were fish finders, Elmer was like a sonar detector on Rockland's inland lakes. Like Josie Kubale was to mushrooms, Elmer was to fish. He would row and row his boat, then gently glide to a spot and instruct his guests to drop his handmade anchors, which were ice cream pails and coffee cans filled with concrete. A recycled metal U-bolt at the top of those anchors allowed him to fasten the rope to the boat.

If the fishing crew missed the spot on the slow roll in, Elmer would row back out to deeper water and try it again.

All those who go fishing have their stories. For me, among the stories that stand out is the day the preteen Corey pulled a snapping turtle into the boat after it took the bait from the end of my bamboo pole. When Grandpa Elmer turned around and saw what I had done, he jumped into action. He cut the line, grabbed the turtle by its shell, and let it glide back into the lake. To this day, I stop and help a turtle trying to traverse a rural road. Lesson from that day? Always keep that snapping head away from your fingers, and you'll be fine.

Then there was the time my father, Randy, allowed Grandpa Elmer to take me and my sister, Angela, out fishing on the day after school let out in spring. Dad, who was a full-time farmer like Grandpa, gave these specific instructions: "Have him [pointing to me] home by 11:00 a.m. We must chop first crop before it rains later today."

That decree came before the sun cracked the horizon, which meant the milk pump would soon be buzzing, as it was milking time. Grandpa always hit the lakes early, even if the eyelids on his grandchildren were barely open.

Well, the fish were biting—so much so that Grandpa and I could not even keep two hooks baited with worms. The fish must have known that the rain was coming, too. Angela, a second grader at the time, just kept dropping the bait in the water and pulling in a fish about ten seconds later.

"Just drop the empty hook in the water and give us a chance to get the fish off the bottom of the boat," advised Grandpa. And that's what she did, God as my witness.

"Grandpa, look, I got another one," she said, smiling as she steadied another pan-sized bluegill in front of his face for another removal from the hook. Grandpa smiled just as big because that fish bit on an empty hook.

Well, we limited out that morning. However, we didn't make it home by 11:00 a.m. Dad was mad. And I knew it as soon as I gazed on his six-foot, three-inch frame. As for me, I had to hide my smile, or I would have been in even bigger trouble.

My internal smile was twofold—I got to go fishing, and I didn't have to scale or gut fish that day, as Dad put me right on the tractor. Grandpa took care of both jobs as Angela looked on.

It's a Wisconsin Tradition

The Friday fish tradition goes back many centuries to a time when the Catholic Church forbid red meat consumption, including poultry, on Fridays. Church theologians wanted the faithful to remember Christ's suffering, passion and crucifixion each Friday, and limiting meat consumption was one of the remembrances just like the great images on church windows. Imagery and reflection were important teaching tools in an era when many of the faithful could not read or write. And since fish was a cold-blooded critter, it could be eaten.

And so, the German, Polish, Czech and other faithful immigrants brought their fish-eating tradition to their adopted Wisconsin. Even the non-Catholics bought into the tradition.

Given the abundance of walleye, perch and bluegill in those days, fish fries popped up at every local tavern. Not only were the freshwater fish good eating, but those Friday night occasions also allowed friends and family to catch up on the local news.

Doubled Down on the Fish Fry

By the time Prohibition became the law of the land from 1920 to 1933, fish fries were deeply entrenched into Wisconsin's culture. While federal law prevented taverns from selling alcohol, fish became the legal bait to draw in business. Just like the "Copper Cows" on the dairy farms in the era, tavern keepers used the fish fries as the front to eventually distribute a little alcohol to wash down the meal later in the evening.

When Prohibition ended in 1933, the Friday Fish Fry had become further engrained in Wisconsin's lexicon and culture. After Vatican II concluded in 1965, Pope Paul VI essentially lifted the red meat edict for Fridays—except during Lent. Even so, Wisconsinites continue to go out in droves for Friday night fish to this very day.

An overwhelming 84 percent majority of respondents to a 2019 University of Wisconsin Extension survey reported going out to a fish fry a few times each year. Another 42 percent shared that they went to a fish fry about once a month. Still, 14 percent stated that they went to a fish fry each and every week.

While every restaurant had its special recipes, Elmer and Julia's children contend the farm kitchen produced the tastiest of fish. What was Julia's secret?

"Mom's fish frying recipe was very simple, at least I think it was," explained youngest daughter Annie. "She always fried in lard, about three-quarters to one-inch deep in the iron skillet. In my opinion, that's the secret to good-tasting fish. Next, she would bread the fish with a combination of flour, beaten eggs, salt and pepper. No thick coating on the fish. That's all. In that way, you could really taste the fish, the wonderful crappy or bluegill flesh, with just a hint of seasoning," Annie went on to explain.

"Once prepared, into the pan the fish went. A quick flip ensured both sides were cooked," added Annie's sister Rosalie.

"I still like my fish that way, and trust me, without frying them in lard they are just not that good," Annie said, as if still in the farm kitchen and hearing the lard crackle on the hot skillet. "One more thing: dad never filleted his fish. He kept the skin on them and the bones in them. That added to the flavor."

"There was no better tasting fish than being cooked the same day as the catch. So fresh," added Rosalie, recalling the local lakes' clarity in those days.

Every Wisconsinite can pinpoint their favorite fish fry. For Annie, Mom and me, that favorite fish fry came from Julia's skillet.

20

LOVE TOOK HIM ON A MICHIGAN ROAD TRIP

What caused a Reedsville farm boy to take his parents' 1952 Chevy farm truck on an unannounced 525-mile road trip?

Young love.

Elmer John "Butch" Pritzl wasn't the first teenage boy to make a choice that made sense to him but agitated his parents—nor will he be the last boy to make such an impulsive decision.

His mother, Julia, must have known something was brewing in the spring of 1960.

She turned to her journal to vent her frustrations.

"Well, this year should have a memorable ring," she wrote on May 10, 1960. "On April 23, Elmer (Dad) put in 13 acres of oats and alfalfa seed. He worked like a demon all day. And Butch lazed around and eventually did chores.

"It was the day after he attended the Lincoln High Prom," she said of Manitowoc High School's signature springtime event. "He came home around 5 a.m. after a prom party at the Armory," she said, noting her son, Butch, went to the prom with Martha Strauss.

TOOK A REGULAR JOB

The word *lazed* stood out in the journal.

To this day, many farm families will have a son or daughter take an off-farm job. It allows them to gain a better appreciation for working for another employer.

Sometimes those young men and women learn the grass isn't greener on the other side of the fence, and they eventually return to the family farm. In other instances, those young people like the off-farm work and never return to the family business.

In the summer of 1960, Butch took a job with Huebner Implement Works in nearby Forest Junction. The farm equipment manufacturer was in its heyday, and Butch had a hankering for all things mechanical and engineering orientated. It was this first off-farm job that convinced Butch to eventually follow his passion, and he went on to become a skilled machinist and model maker for a number of Two Rivers, Wisconsin machine shops.

Back to the summer of 1960—to help get the farm work done on the Pritzl Farm, Butch's older sister, Jacque, came home from Ripon College.

"This summer, Jacque is staying home and working like a demon," wrote Julia. "The work is hard for her.

"Butch works at Huebner's in Forest Junction and makes $1.05 per hour. He works from 8 to 5 p.m.," she said, noting that he carpooled with his friend Ron to cut down on the transportation expense.

Just Going to the Bank

"Butch got his payday and went to the bank in the evening," wrote Julia, noting that he went to cash his paycheck at the nearby Reedsville State Bank. However, he never came home that Friday, July 1, 1960. Elmer and Julia must have been worried sick.

Butch eventually called his parents the next day from across Lake Michigan—as in the state of Michigan.

"Instead of coming home from the bank, he drove in our farm truck to Muskegon, Michigan," Julia wrote, underlining farm truck for added emphasis. Not only did Elmer John not come home, but he also took the farm's main transportation for hauling supplies.

"That's 525 miles," she wrote, most assuredly having taken out a map and calculated the distance.

It was Butch's desire to see Martha, who was living in Muskegon, Michigan, for the summer that fueled his decision to take the road trip. Martha's father, Harvey, was a mariner who spent his entire life sailing on the Great Lakes. The family had properties in both Manitowoc, Wisconsin, and Muskegon, Michigan. The two cities were "bridged" by a car ferry service that "connected" U.S. Highway 10 across the Great Lake.

Above: While Elmer and his son, Elmer John, got along quite well throughout life, the Pritzl parents decided to send their two younger daughters on a "vacation" to Aunt Beatie and Uncle Quiren Sleger's home to sort out their son's Michigan road trip. *Author's collection*.

Opposite: Heading into his senior of high school, Elmer John "Butch" Pritzl made a daring dash to Michigan to visit his girlfriend. He made that dash in the Pritzls' farm truck—a green 1952 Chevy. *C. Todd Garrett*.

"May the Lord take care of him!" Julia continued in her journal, clearly writing as the events unfolded day by day.

Not much was said of the reaction of Butch's father, Elmer.

However, the next general entry may have laid the notion that an adult-type conversation was forthcoming.

"The girls are vacationing at Beatie's for five days and left on Sunday evening. We will get them Friday or so," wrote Julia of the July 3, 1960 decision. The *so* may have been the number of days needed by Butch's parents to sort this out. The sorting part didn't require daughters Rosalie and Annette, who were ages eleven and eight, respectively.

MAKING AMENDS

That Friday, July 1 trip was quite the ordeal, and it surely went deep into the morning hours of July 2. Butch drove through Michigan's Upper Peninsula and down to mainland Michigan. From there, he trekked to Muskegon on Lake Michigan's eastern shore.

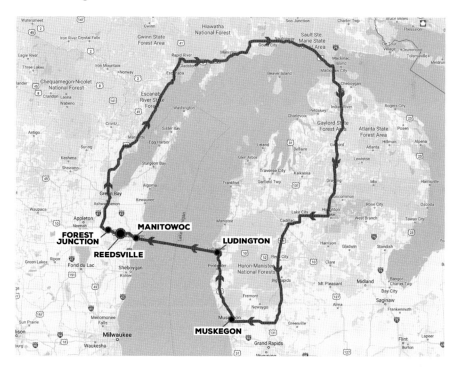

Why make the drive instead of a ferry voyage?

By the time Butch completed his Friday workday, there was no time to make the car ferry out of Manitowoc. So, he drove all the way around the lake.

Once in Muskegon, Butch got to spend the Fourth of July weekend of 1960 with his girlfriend Martha.

But even the near-cosmic power of young love could not keep Butch from the reality that he eventually needed to get the farm truck home and face the wrath from his German father, Elmer.

To expedite the trip home, Butch took the car ferry from Muskegon to Manitowoc. During the four-plus-hour boat ride, a few passengers learned of Butch's weekend escapades, and some of them volunteered to help wash the farm truck. So, a small group of people went down to the automobile hold, where they washed and waxed the truck, thinking it might soften the ramifications from Butch's parents when he returned to Reedsville.

While no more was ever written on the matter of the Muskegon road trip, Martha and Butch eventually married and became parents of three wonderful children. In the subsequent years, short innuendos were made about that daring road trip, the 1952 Chevy farm truck and escapades to Michigan.

Secrets can be kept, for a time, but eventually the full story sees the light of day.

21

FLING THAT RING

Please tell me you didn't haul out the manure!" Julia exclaimed to her husband, Elmer.

It's not a statement one often hears—even in dairy farming circles. That surely holds true when it comes to a conversation regarding daily manure hauling from a dairy barn in the dead cold of January.

In normal circumstances, Julia would have been simply happy to hear that Elmer got the tractor started and was able to haul that manure without any breakdowns—or worse, getting stuck in the snow with a two-wheel drive tractor that had the slight benefit of chains on its rear tires. To be candid, manure hauling was a dreadful winter chore.

Anyone who has hauled manure in an open cab tractor in January knows firsthand how the cold winds rip at one's face. A seasoned operator also knows the tractor driver had to go into the wind—as opposed to with the wind—so that the manure didn't blow back on you once the beaters flung it in the air.

Once the manure was out of the spreader, the hauler still had to get off the tractor and scrape the spreader by hand. That scraping action with a metal scraper was critical, or the warm manure would eventually freeze onto the spreader, and nothing would work the next day.

Clearly, manure hauling was a "last on the list" project. As Elmer looked on his wife after her plea, he could sense the angst in her voice.

Julia was a practical woman who lived life with very little drama. She was a hardworking farm wife who milked forty cows side by side with her

Elmer, standing in the milk house, almost cracked a smile as to declare he was done hauling the manure on the cold January day. Moments later, the smile vanished as he learned his wife's ring was flung out with the manure. *Author's collection.*

husband morning and night. But at this particular moment, she looked as if she was going to start crying in the milk house, where the milk from the family's dairy herd was stored in the bulk tank cooler prior to the truck picking it up and hauling it to the processing plant.

Julia also was slightly breathless, having just run from their farmhouse bundled in her winter clothes.

Elmer was puzzled.

"I just got done," said Elmer, almost cracking a smile. "The tractor started on the first try this morning," he added with a hint of relief.

"Oh, no!" exclaimed Julia, who now had distress oozing from her voice and a facial reaction that matched.

Elmer became confused.

"My ring…it was in the barn gutter!" confessed Julia. "It got mixed with the food waste when I was washing dishes and I dumped it."

"Oh no!" Elmer exclaimed, interrupting his wife. Elmer was beginning to understand Julia's distress. Elmer's anxiety had spiked and may even have begun to outpace Julia's angst. Elmer knew what happened—it wasn't pretty.

Julia's wedding ring, the one he gave to her twenty-five years earlier on June 16, 1938, had been hauled out with the cow manure that morning. Who knows where that manure spreader flung that diamond and its gold wedding band.

A WINTER TO FORGET

Losing a ring just topped off what had become a dreadful winter. That particular January was nasty cold.

"Winter of 1962–63 was one of the coldest winters in history," Julia wrote in her farm diary. "March 8, we have our 49th day of below zero weather. All January was below zero.

"Elmer is 'fed up' with wrestling the elements. Tractor didn't want to start all winter, so he put it in the new stable for a long time," wrote Julia of their ninety-six-year-old family farm. "He began to grow scared of fire, so I told him to put it in the shed," she continued, with the fire danger being a factor if sparks came out of the muffler.

LOST IN A SINK OF DISHES

On dairy farms, it was common practice to save food scraps from the family dinner table, place them in a small bucket and either feed them to hogs or dump those scraps in the barn gutter that held cow manure. If dumped in the gutter, eventually, those table scraps would end up on a field, decompose and become nourishment in the form of fertilizer for future crops.

However, wedding rings don't exactly decompose. And a rock and its metal band surely should not end up as fertilizer on farm fields.

Earlier that January morning, Julia had helped Elmer milk the cows. On the way to the barn, she carried with her a pail of table scraps and dumped them in the gutter. It was a daily routine, and she might have done it sleepwalking on her 5:00 a.m. journey to milk the cows.

On their twenty-fifth wedding anniversary in 1963, Elmer presented Julia with a new wedding ring to replace the one that was flung into the farm field by a manure spreader. *Author's collection.*

Once the couple was done milking cows, Julia went to the house to make breakfast. Elmer continued with chores, which included hauling manure.

When making breakfast, Julia started looking for her wedding ring, as she liked to wear it in the house.

She looked and she looked.

That's when she got a knot in the pit of her stomach. "I lost my wedding ring in the sink full of dishes, washed them, and forgot about it when I pulled the screen for the drain," she wrote in her farm diary.

Julia quickly bundled up and sprinted toward the barn.

Elmer and Julie trudged out into the field and began a hopeless search for Julia's diamond wedding ring. Ironically, that ring was spread somewhere in a field where Julia's grandparents had built the family's original log cabin on arriving from Bohemia in 1867. With manure flung everywhere on the January snow, the couple did not look long—finding it was simply a lost cause.

"Julia, I'll buy you a new one," Elmer said to his bride of twenty-five years.

"I'm so sorry, Elmer," she said, crying to her husband.

Five months later, the couple's older children, Jacqueline and Butch, planned a dinner party with dancing at Dill's English Lake. Those two adult children each chipped in $24 for the meal. Elmer caught the drinks at $49—that bill was $1 more than the food.

And Julia received her new wedding ring that night, too. Elmer purchased it from Fehr's Jewelry Store in Manitowoc. Coincidentally, that very store was built the same year Julia's grandparents built their first log cabin in the nearby farm field in 1867, where the first ring lies to this very day.

22

THREE LONGS AND A SHORT

'd like to place a call to Wisconsin's Rockland 1731," said Bob Krueger.

"Sonny, I have no idea what you're talking about. I need a seven-digit code," replied the telephone operator based in New York City.

"The exchange does not have a seven-digit code. Just put me through to Rockland 1731 ring dial number. Its three longs and a short," replied Bob, who was trying to complete a call from a relative's home in New York City to his sweetie in Wisconsin's Manitowoc County's Rockland Township. The three longs and a short was code in the day for 1731.

Now the New York City operator thought the young man on the other end of the line was a bit dim. "I don't think that's possible," stated the operator. "I must have a seven-digit number," she repeated. "Telephone lines these days require a seven-digit code."

"Not this one," countered Bob in late July 1969.

He knew better, because his home Wayside and the Rockland exchanges were among the last holdouts throughout Wisconsin, as both were still using manual operators that pulled cords and connected calls on switchboards. Now Bob really wanted to talk to his girlfriend Annette "Annie" Pritzl, who was about to be maid of honor in her big sister's wedding on August 1. Bob could not make the wedding.

That's when he came up with another plan. Bob's idea? Have the New York City operator talk to a Wisconsin operator. At first, the New York City operator talked to a counterpart in Milwaukee, Wisconsin. That wasn't going too well.

When Bob Krueger called his girlfriend Annie Pritzl to share he could not make it home for her sister Rosalie's wedding, the call was near impossible to complete, as the Rockland Telephone Company still used open wire and manual mechanism to make telephone calls. Pictured here are (*left to right*): Rosalie, Elmer John, Annie, Elmer, Julia and Jacque. *Author's collection.*

That's when Bob dug a little deeper.

"Can you put me through to the Appleton, Wisconsin operator?" Bob asked during the late-evening call, with the idea that getting closer to the Rockland area might actually make the call happen.

"I can do that," said the operator working out of the Big Apple.

Bob explained to the Appleton operator that he was trying to call the Rockland exchange at 1731.

"I'll help the New York operator connect the call," said the lady based in Appleton. "I'll bet she can't believe there are still party lines," she said to Bob with a chuckle.

And so, a conversation ensued between the New York and Appleton operators well before the days when area codes were created that would further extend phone numbers from seven to ten digits. The New York City operator eventually came back on the line: "We can now connect your call, sir. Have a good evening."

With that twenty-minute prelude out of the way, Bob was able to talk to Annie.

Bob's cousin Rose, located in New York City, could see that Bob was in love. "Keep talking—I'll pay for the call," she said to her younger cousin.

THE LAST OF AN ERA

Area legend has it that the Rockland Telephone Company was Wisconsin's very last telephone network to use open wire and manual mechanisms to make telephone calls. Two documentaries would deem that legend to be true. On December 29, 1967, WTMJ television out of Milwaukee trekked up to the rural Rockland home of Ralph and Florence Moede. The footage is now archived by the University of Wisconsin–Milwaukee. Its archive summary states, "The Rockland Telephone Company, Inc., is based in the home of Mr. and Mrs. Ralph Moede of (rural) Reedsville. The company used open wire and old-fashioned manual instruments and fast disappearing party-line system. The children also help."

Believe it or not, WTMJ Channel 4 filmed that segment in color, not black and white. The irony? The television station was filming a story in color about technology that was still being used prior to motion pictures becoming mainstream.

That documentary was filmed in 1967, and the same technology was still being used a few years later to place that aforementioned call between Bob Krueger and Annie Pritzl.

In the August 20, 1970 edition of the *Herald Times Reporter* based in Manitowoc, Wisconsin, the newspaper did indeed declare that the Rockland Telephone Company was the last rural telephone exchange using a magneto switchboard. The very next day, that switchboard went silent.

In those days, it cost a family $2.75 per month to have a home phone on the Rockland Exchange. Payments were due on a quarterly basis. Once direct dialing went into effect, the cost jumped to $8.25 per month for the base rate. To stabilize the service, all the telephone lines were buried underground, as opposed to being strung on poles.

SLOW TO THE COUNTRY

Living in the country a century ago came with some technological baggage. Electricity was slow to make its way to rural farmsteads, as were telephones.

While Rockland continued to use a magneto switchboard until 1970, adjacent Reedsville switched to dial phones in 1955. Shown here is the final days of the open wire and manual mechanism switchboard in the village. *Village of Reedsville.*

It's the reason city cousins often referred to their rural classmates as "country kids" or "country bumpkins." Those happen to be a few of the printable names. Living in the villages, towns and cities made those kids a bit more street savvy. In some ways, the lack of high-speed rural broadband is a modern-day version of this technological blackout legacy.

In the case of telephones, nearby Reedsville had a three-year head start on its country cousins, as the village had telephones beginning in 1905. That technology came courtesy of the Manitowoc and Western Telephone Company. Three years later, three men sought to bring telephones to the Rockland Township and close the technology gap.

"The Rockland Telephone Company has filed articles of incorporation with a capital stock of $41,500. The incorporators are Albert Moede, Albert Reinke, and Ferdinand Kleist," recorded the publication *Commercial West* in its Saturday, March 14, 1908 edition. Albert was the grandfather of Ralph Moede, who later would run the same company. That same publication shared that the Bell Telephone Company would expend $700,000 on improvements to its system in Marshfield, Wisconsin.

A Family Affair

Most assuredly, the Rockland Telephone Company was a family business. Albert Fischer served as manager of the Rockland Telephone Company for fifteen years. Fischer, who was born in Germany on August 6, 1864, happened to be the brother of Wilhemina Reinke, wife of one of the three founders.

Richard Moede served as a lineman for the Rockland Telephone Company for fifteen years. That meant Richard was out setting telephone poles and stringing telephone lines to make connections between each residence and the main switchboard hub. That also meant Richard had to fix telephone poles in the event of storm damage. While there is no written documentation of storm damage to the Rockland Telephone Company, on February 22, 1922, a sleet storm took all the telephone poles and wires down in nearby Reedsville. This resulted in no service from February 22 to June 12, 1922. Due to the proximity, Richard likely was making repairs, too.

Richard was an interesting fellow. He was a fur rancher for fifty-three years and even was a Manitowoc County deputy sheriff. Richard was the son of Albert Moede, who served as the first president of the Rockland Telephone Company.

Richard was also the father of Ralph Moede, who was the third generation to run the company. Ralph's wife, Florence, was a telephone operator in her in-laws' home for ten years and then in her own home for seventeen.

Ralph's mother, Ruth, also served as an operator for the Rockland Telephone Company from 1939 to October 1953. She died one year later.

One person who didn't appear to have any family connections to the founders, William H. Rusch Sr., served as treasurer of the company for several years. That's the same Rusch family that ran the Reedsville's bee business.

THE COMPANY'S FINAL DAYS

On August 28, 1969, the New Rockland Cooperative Telephone Company, the organization's official name, placed a formal application with the Public Service Commission of Wisconsin to sell its assets to the nearby Valders Telephone Company. In those days, Rockland had 142 company-owned stations. That meant Rockland had 142 hand-crank phones in residences. Meanwhile, Valders was a much larger organization, with 1,412 company-owned stations throughout the Collins, St. Nazianz and Valders area.

There was some opposition to the sale.

There were eighteen Rockland customers northeast of Reedsville, sixteen within Reedsville itself and one located on U.S. Highway 10. In 1929, the Railroad Commission of Wisconsin allowed those stations due to special considerations for farm residences that did business with the Reedsville Cooperative Association. In short, being on the Rockland Exchange made their calls free to the co-op. The Railroad Commission had jurisdiction at that time because telephones were the upgrade to telegraphs that often ran along railway tracks.

In the early 1900s, these phones used by the Crandon Telephone Company, along with the likes of the Rockland, Reedsville and Wayside telephone companies, represented the latest in technology. *Charles Wilbur.*

In 1969, the Public Service Commission of Wisconsin said no more. The Valders Telephone Company would have to divest itself of those thirty-five residential stations.

Just a few years after the sale, Valders would do away with all the crank phones and switch to dial phones. Those were the same dial phones that had come to nearby Reedsville in 1955.

Meanwhile, the other holdout using the party-line system, Wayside, had made the leap from hand-crank phones straight to the state-of-the-art push-button phones. In doing so, that telephone network skipped right past the dial phones that would be headed to museums in a few decades.

Phones Were for Business

During its entire existence, the Rockland Telephone Company operated under the party line system. That means if a person on your exchange was on a call, you could listen to the entire conversation.

That also meant if a person on your exchange was on a call, you could not make a call yourself. Essentially, you had to wait your turn.

That's why many adults considered the phone a necessity of business, not a mode to catch up on social conversation—even though on some occasions people on the same exchange would have long-winded discussions.

"One time Dad dashed into the house," said daughter Rosalie of her father, Elmer. "He ran straight to the phone. When he picked up the receiver and heard two women talking about laundry, he was furious."

"Ladies, get off the phone. This is an emergency. I need to call Doc Junge!" he bellowed out. "I have a cow that's having a real tough time having a calf."

"The ladies quickly hung up, and Dad was able to make the call," recalled Rosalie.

There was another reason for the "phones are for business" mindset—calls outside your exchange were long distance and the "call meter" ran by the minute.

"It cost me a nickel to call home from college," said Rosalie of calling her parents from Stevens Point. "If the call went too long, the operator would come on the line to tell you to put in more money," she said of calling from a phone booth in the university's dormitory.

It's also the reason you see people in old films with small piles of coins. If they were making a long-distance call, that person had to be ready to place more money into the pay phone.

"That's why you didn't use the phone in those days like you do today," she continued. Rosalie Geiger and her brother-in-law Bob Krueger were among the last Wisconsin residents to say "three longs and a short" to call 1731.

23

THE BIG ELMS FELL

The American elm—it's one of North America's grandest trees.

Indeed, the mighty elm has witnessed multitudes of historic events—perhaps none more transformative than that which took place in the shadows of the "Old Washington Elm" located in Cambridge, Massachusetts. It was under this venerable American elm that George Washington was named a general by the Continental Congress. With that commission, General George Washington took charge of the Revolutionary troops.

The world was never the same.

As America's cities grew, it was the American elm that began lining their streets and boulevards. There were plenty of good reasons.

Elms had grand, picturesque canopies.

Elms were hearty trees.

Elms were tolerant of pollution.

Elms thrived in compacted soil.

Elms could even withstand road salt.

And so, the American elm became even more deeply entrenched into Americana. Those in older generations fondly recall its distinctive vase-shaped canopy cascading from its sturdy trunk. The American elm went on to become one of the most widely planted shade trees in the United States. Nearly every village, town and city has an Elm Street.

Where have all the elms gone?

THERE'S NOTHING DUTCH ABOUT IT

Dutch elm disease, as it's now called, arrived in America in the 1930s. Ironically, there's nothing Dutch about this disease. Global trade brings global diseases.

Dutch elm disease was first described in the Netherlands in 1919. However, that's not where the disease originated. By all reliable historic accounts, the disease likely arrived in Europe via Asia during World War I. The disease spread rapidly and by the 1930s could be found wreaking havoc in all the countries of Europe.

Dutch elm disease was accidently unleashed in the United States in northeast Ohio. In 1929, a steam train stopped for coal and water in Creston, Ohio. Its manifest included a shipment of veneer elm logs from France. The next year, elm trees along the train tracks started to die, and Dutch elm disease quickly spread to nearby Cleveland, according to reports by the Ohio Chapter of the International Society of Arboriculture.

By 1933, elm trees around the Port of New York City were beginning to die. Fast forward to the present, and Dutch elm disease has claimed over 100 million American elms.

BEETLES AND A FUNGUS

Dutch elm disease is largely caused by two different fungi. However, fungi aren't necessarily that mobile in their own right. Enter the European elm bark beetle and the American elm bark beetle.

The female beetles seek out places to lay their eggs. A typical landing spot is a V-shaped entry point where a branch connects with a larger branch or the trunk of the tree. It's here that the female beetle excavates an egg-laying gallery underneath the tree's bark.

If the fungus is present, spores proliferate under these egg nests. When young adult beetles emerge through the bark, many of these insects carry the spores on their bodies. As the reproductive cycle begins anew, the next generation of beetles goes forth seeking new nests and subsequently carries the disease to more elm trees.

As for the death of the tree itself, some fungal spores make their way to the water-conducting vessels known as the xylem. The fungus begins to block the tree's ability to transfer water internally, and the tree begins to die. The disease is so communicable that healthy elm trees can contract the disease if its roots are intertwined with a diseased tree.

Young elm trees die rather quickly, while it may take a year or more for large, stately trees to meet their final demise. In either instance, the trademarks of the disease's onset are unmistakable.

At first, the elm's upper branches wilt and leaves turn dull green to yellow and curl. Eventually, those leaves become dry, brittle and brown. This condition then cascades throughout the tree until all the leaves are gone.

Wisconsin's Elms Fall

Even though Ohio's elms began dying in the 1930s, it wasn't until the 1960s and early 1970s that the Badger State's elms began to die. But once the disease arrived, it was swift moving.

While Elmer and Julia Pritzl didn't live on Elm Street, their farm did have an "Elm Woods." Many farmsteads would name parcels to identify areas of the farm property.

The years 1972 and 1973 were the beginning of the end for a vast majority of the area's elm trees. The Pritzls' Elm Woods transformed from shades of green, to yellow to brown—and it wasn't even August.

Old-timers recall that American elms once lined Jerabek Lane. The graceful trees once swooped over the gravel road, creating postcard-like images. By 1972, wooden skeletons stood along that road and the entirety of the Pritzls' Elm Woods.

The same story unfolded as if winds had carried a wildfire across the countryside. In this case, however, the wildfire was carried by beetles and the flames were a fungus.

"An avid arborist and woodsman, Elmer began cutting the elms down at a feverish pace," recalled Steve Reis, who worked on the Pritzls' farm from 1973 to 1975.

The tree trunks were loaded on wagons and hauled to a nearby mill to be cut into lumber. The treetops became firewood for the family's home. And everyone involved in the process expended a lot of energy because elm is a mighty tough wood. The wiry wood is hard to split and puts up a good fight before succumbing to a multitude of swings from an axe or maul.

Once lumber returned from the mill, the ever-frugal Elmer piled the lumber on his farm with the help of Steve, placing space boards in between each layer to allow air to dry the boards. Six months later, the piles were

For decades, Elmer hauled large logs to nearby mills for later use as boards on the farm. Seated on the logs are daughters Annie and Rosalie. *Author's collection.*

deconstructed, the boards flipped and piled once again. In this process, Elmer could avoid paying the kiln-drying fees.

In 1987, as a high school freshman, I took some of that elm and built a flat rack wagon to transport hay. Once completed, it measured sixteen feet long and eight feet across.

The only reason Winton Jaeger even permitted a freshman to undertake a project reserved for a high school senior is that the Reedsville High School's agriculture instructor with three-plus decades of teaching experience knew Elmer would be keeping a watchful eye on his wide-eyed grandson, who tended to tackle projects too big for his britches.

The flat rack made of elm-planked lumber lasted a few decades. Elm, as it turns out, might be tough to split, but when left out in the weather, it rots mighty fast. Like the American elm, that flat rack, even though its boards were painted on both sides, eventually succumbed to Mother Nature.

THE TWO-AND-A-HALF-HOUR CUT

By the time August 1972 rolled around, Elmer had plenty of experience cutting down trees on his farm and nearby roadways. That's when he became bold enough to agree to drop a dead elm nestled on a property at 310 Cleveland Street in Brillion, Wisconsin.

Elmer arrived upon the scene early in the afternoon on August 13, 1972. He came to work wearing his customary blue jeans, denim shirt and denim cap. After rolling a cigarette, he carefully surveyed the situation. One tree specialist quoted $300 to take that tree down. Elmer thought he could master the situation for his daughter Jacque and her husband, Ken Konop, who had purchased the 310 Cleveland Street property in June 1969.

What did he want in return?

Help from his young son-in-law making hay.

"That was a big event!" recalled Ken. "The tree was huge and grand."

After laboring for two and a half hours, Elmer posed triumphantly on this stump after felling this American elm that stood nearly four feet in diameter. Dropping a tree like this in the woods is one feat, but sticking a perfect landing between two homes was like landing a large military aircraft on a municipal airport runway. Elmer stuck a perfect ten landing. *Author's collection.*

"It was so sad to take down such a big tree. However, it was exciting to see, as it was a dramatic and risky procedure. How that tree didn't crush our home or Zanders' next door is just impressive," Ken's daughter Julie went on, recalling the events of that day.

"At 1:30 p.m. Elmer made his first cut into the tree," wrote Julia in her journal. "That tree was 11-1/2 feet in circumference," she continued—the tree was nearly four feet across at the base.

That tree was indeed a giant.

For the next two and a half hours, Elmer notched the tree and continued to survey the situation and apply every ounce of his wisdom to the matter at hand.

As four o'clock approached, Elmer alerted the dozens of bystanders that the time for the final cut had come, and he urged the band of onlookers to move away. Then, he began to make his final cut with his powerful Homelite chainsaw, moving from the back end of the tree's trunk and making his way toward the notched area.

"Timber," he yelled, with beads of sweat pouring down his brow and face in the hot August sun.

The American elm at 310 Cleveland Street came crashing to the ground.

A perfect landing. The Konop and nearby Zander homes didn't have a scratch. One of the onlookers brought Elmer a shot of whiskey. While he wasn't much of a drinker, Elmer thought a second one was warranted that day. One onlooker gladly poured it for Elmer, who would turn fifty-five years old two days later.

"The tree was on the ground for a few weeks afterward," recalled Ken. Said Julie, "We played house in its branches….The branches provided an instant structure to crawl around in."

Ken and Jacque slowly cleaned up the tree, using Elmer's hay wagon that they also used to make hay that summer.

Sadly, the American elm's fateful journey is being replayed in another tree species.

"Interestingly enough, in early June 2020 we took down four big ash trees in our yard," said Chad Konop, the younger brother of Julie. "Indeed, the emerald ash borer is the modern-day equivalent of Dutch elm disease."

24

SHE LOVED HER SNOWMOS

She had a need for speed.

She lived out that need for speed as an adult when she became a snowmobile racer in the United States Snowmobile Association, simply known to racing fans as USSA.

Born Jacqueline Ann Pritzl, "Jacque," as she was known to her friends and family, was the eldest child of Elmer and Julia Pritzl. Ever since entering the world on May 30, 1940, Jacque had been daddy's favorite, and her two younger sisters knew it, too.

She also was Grandma Anna Burich's favorite. Even as a child, Jacque had an inherent ability to relate to adults, and she got along swimmingly with Grandma Anna, who was born way back in 1877.

"I am a little bit lonesome for Jacque," wrote Grandma Anna, who was spending some time at another daughter's house. That happened to be a rarity, because Anna and her husband constructed the family's farmhouse in 1916 and she was a permanent resident until she suffered a stroke at the kitchen table in 1951. "Please give my best regards and a couple of kisses to her from Grandma," she continued, writing in perfect cursive.

That was among the one hundred or so words Anna penned to her daughter Julia on September 19, 1944, in what was dubbed "the only letter I ever got from my mother," as Julia wrote in ballpoint pen in her diary. She often did this to ensure those who followed knew all the details.

DUST PLUMES BILLOWED

Back to the need for speed: throughout high school, Jacque drove her younger brother and two sisters to school in the family's Nash Rambler. During dry spring or fall days, Jacque could make that Nash roll like a race car, and Manitowoc Road was her speedway.

"When coming home from school on Manitowoc Road, a dust plume would sail high into the sky as we flew home on the two-mile stretch of the gravel road," said younger sister Rosalie, who knew Jacque was going well beyond any advised speed. It was "Advised" because Wisconsin's rural roads didn't have a posted speed until 1973.

"None of us said a word to Mom or Dad either. Jacque ruled the roost," said Rosalie, who was nine years younger.

Jacque insisted on excellence and thought it her duty to teach her younger sisters German and make them read the Bible. It's as if they got an additional lesson before and after school each day.

Jacque always drove hard.

She not only drove cars hard, but she also studied hard and worked hard on the farm. The nose-to-the-grindstone attitude on the farm gained her favor with her workaholic father, Elmer, and the classroom work that eventually led Jacque to be named salutatorian of the Reedsville class of 1958 made her studious mother, Julia, glean with pride.

Jacqueline Pritzl would be the first in her family to go to college. In choosing Ripon College, she went on to study chemistry and German. *Author's collection.*

Jacque mastered many things—except the clarinet. Whenever younger sister Annie hears a "squeak" from a clarinet, she still thinks of her sister Jacque.

Always in tune with current events, Jacque had a towering white pet cat she named Khrushchev. It was the height of the Cold War. And that cat was named after the pasty-white Premier Khrushchev himself, who would later joust with President Kennedy during the Cuban Missile Crisis.

In many ways, Jacque was the female equivalent of Elmer, who was pure German to his very core. There wasn't a problem she couldn't solve, and if she thought you wronged her, you better watch out because it wouldn't happen a second time. "Wrong me once, shame on you. Wrong me twice, shame on me," Elmer would often say, and his Jacque lived by that motto, too.

When Jacque wanted to enroll at Ripon College, her father, Elmer, just worked harder and came up with the money. She returned the gesture by laboring on the farm during her summer breaks. It's as if Elmer and Jacque were trying to outwork each other.

The Dutiful Granddaughter

Right after graduating from Ripon College, Jacque married Ken Konop on August 4, 1962. The dutiful granddaughter honored her deceased Grandma Anna by wearing her veil on her wedding day.

Jacque always dotted her i's and crossed her t's. And that wedding veil honored her dear Grandma even though she was no longer walking this earth.

After a short stint working for the Institute of Paper Chemistry in nearby Appleton, Wisconsin, Jacque went on to teach at Valders High School for one year. In 1964, she began teaching at Brillion High School. In the spring of 1969, she finally earned her master's degree in chemistry from Platteville State College.

Ready for New Challenges

Jacque was always ready for the next project. In the same month that she earned her master's degree, the Brillion Education Association selected Jacque as its president-elect. She was doing all this even as she had two little children—a daughter, Julie, and a son, Chad. Her teaching career was blossoming, and others were taking note.

The Brillion Education Association named Mrs. Konop its "Teacher of the Year" in 1972. By then, she had gone on to teach chemistry, biochemistry, physics and German. She also chaired the school's science department, was advisor of the German and science clubs, coached the forensics team and organized Brillion High's very first pom-pom team.

The inspiration for that pom-pom team came from her kid sister Annie. As a high school junior, Annie helped form the Reedsville High School pom-pom team with the school's physical education teacher. Annie went on to become the squad's first captain. After a sporting event with archrival Brillion, Jacque picked up on the idea. Two weeks later—poof—Brillion

High School had a pom-pom squad. Jacque was a competitor, even with her very own family!

Her husband, Ken, had also become a steady influence on students, as he was serving as Reedsville High School's principal.

When Jacque was named the teacher of the year, the local news article also listed her new hobby—snowmobile racing.

"It was Ken who got me into racing, but he also taught me to do some of my own mechanical work," she said in a December 1, 1972 interview with the *Post-Crescent*. "I have 'torn down' both his and my machine for inspections. They do run again when I put them back together, too!"

She added that statement because most men in the day would never think a woman could be mechanically minded—but Jacque was a quick study. Plus, she had two great teachers: early farm mechanics from her father, Elmer, and snowmobile mechanics from her husband, Ken.

In those days, the USSA had a mandatory teardown requirement for the top three places at its major races. If that sled was found in violation of the association's rules, the winner was disqualified. Since Jacque had started to do a lot of winning, she also was becoming a wiz at tearing down and rebuilding her sled.

As if teaching had become an afterthought to snowmobile racing, the *Post-Crescent* writer added this narrative from Jacque on the subject of teaching, "I find great satisfaction in teaching, yet find myself constantly inquiring whether the job could not be done better. It hurts when you can't seem to reach someone or explain something thoroughly," stated Jacque.

To the Winner's Circle

Jacque's teaching and sledding careers were taking off. In January 1973, she took first place in the Mod I competition in Weyauwega, Wisconsin, on her Polaris machine sponsored by Potter Sports Equipment. A few weeks later, she finished second in a race in Neenah. That same weekend, Ken finished first in Crivitz.

By March of that year, both Jacque and Ken had qualified for the invitation-only World Series of Snowmobiling at Malone, New York, just fifty-five miles south of Montreal, Canada. Ken piloted a Polaris Starfire factory special with a 340cc engine, while Jacque had a 295cc engine.

The world championships featured the top ten drivers from each of the four USSA divisions, the ten Canadian drivers and the top ten from

Potter Sports Equipment sponsored Ken and Jacque Konop, who are shown on their racing sleds. *Author's collection.*

the American Association of Drivers. Ken had finished sixth in the USSA's Central Division with 703 points.

"Jacque Konop, a pretty brunette, ranks fourth in the division with 221 points," wrote a reporter for Appleton's *Post-Crescent.* "She will race in Women's Mod-I class and qualified last year in 'A' stock." *Mod* was short for modified.

The World Series competition knew the couple could rank higher in the standings.

Teaching did rank higher than snowmobiles for Ken and Jacque, but not by much. Both had competed in Fond du Lac, Milwaukee, Shawano, Antigo, Neenah, Weyauwega and Houghton. However, some races presented a problem, such as the Eagle River affair, which involved three days of racing. That event conflicted with their teaching profession, and both bowed out of the race, losing out on any potential points to propel them up in the standings.

Up until that point in the season, the Konops had collectively recorded more than 5,800 miles going to races and spent just over $1,000 on entry fees, $3,000 for the two machines and another $500 in parts. The racing season began on December 1 and drew to a grand finale with the World Series event.

"This is the ultimate in racing," said Ken to a reporter for Manitowoc's *Herald-Times Reporter* prior to leaving for the race in New York. "There's no easy draw in these races. You just hope that you don't bump head on into four division champions in the heat races."

Ken had qualified for every world championship since the events began in 1970. He won the World's Championship in 1971 in the "C" stock classification. In that interview, Ken noted he had a few injuries, although they had been minor.

For the Konop family, that injury record was about to change.

25

HE FELL TO HIS KNEES

The snowmobile season of 1974 picked up right where 1973 had left off. On January 19, Jacque Konop posted the fastest time at the USSA race in Clintonville. Two weeks later, Jacque powered her Polaris Starfire bearing race number "1386" to another top finish at the Antigo series race.

Next on the docket was the Chilton race, and some Green Bay–area television news crews were headed to that February 9 event to cover the rapidly ascending Jacque Konop. She now ranked in the top fifteen among all women in North America in the USSA standings.

Jacque was a rising star on the racing team now sponsored by the Pabst Brewing Company of Milwaukee, Wisconsin. In those days, Pabst ranked right up there as a leading brand with Budweiser and Miller.

AND THEY'RE OFF

Jacque took an early lead out of the gates at the Calumet County Fairgrounds. That's when everything went wrong.

Having just completed her first lap in the women's "Mod I division," she started to make her second lap around the oval. She hit the throttle as she was coming out of the first curve and headed into a straightaway. That's when Jacque's sled flipped, she was thrown and then the sled fell on top of her.

Jacque Konop had hit the big time and was sponsored by one of America's largest brewers—Pabst Blue Ribbon. *Author's collection.*

No others had been involved in the mishap.

The crowd fell silent.

It was 3:45 on that Saturday afternoon.

Medics quickly rushed to the scene, and they didn't like what they saw.

Jacque was unresponsive.

An ambulance rushed Jacque to the nearby Calumet Memorial Hospital minutes later. An X-ray revealed that she had a broken neck and possible spinal cord damage. The doctors conferred. If they could save her, she still might never walk again.

Regrettably, doctors never got to discuss the "walk again" prognosis, as Jacque died within the hour. The sled not only broke her neck, but the impact also collapsed both of her lungs.

NEXT OF KIN

It was mass chaos everywhere.

Since Jacque's husband, Ken, was there that fateful day and several media crews were on-site, the local authorities gave the all-clear and released the name of the deceased.

Snowmobile racer Jacqueline Konop was dead at age thirty-three.

Television and radio crews began airing the reports at the top of the six o'clock news that Saturday night. It hadn't even been two hours since the Calumet County coroner had pronounced Jacque dead.

During the chaos, Ken asked for close friends to call Jacque's siblings. In turn, they could deliver the dreadful news to her parents.

The Chilton race officials, having received word of Jacque's death, held the quickest vote in race history and immediately awarded its A.J. Horst Memorial President's Trophy to the Konop family.

The day turned into a blur—a nightmare, a bad dream with multiple chapters yet to unfold.

Back at the Farm

Jacque's sister Rosalie had been planning to go to the Chilton races that day. She'd never seen her big sister race before. For Rosalie, it was the closest race of the season to her home in Calumet County.

But a sensation came over Jacque's younger sister that very morning before she left. She couldn't pin down her feeling. She decided not to go.

Just as Rosalie and Randy were finishing up dinner and were about to milk cows on their rural Brillion farm that night, the phone rang. Rosalie picked it up.

"Is Randy there?" the caller asked. She recognized the voice as Father Raymond Dowling, the parish priest at St. Mary's Brillion, who had baptized her son a little over a year ago.

"He's heading out the door. Can I help you?" asked Rosalie, who saw Randy had his barn clothes on and was ready for work.

"Please get him. I really need to talk to him," said Father Dowling, who was beyond persistent. Rosalie was frustrated; she didn't know why she couldn't take a message. Some men were just old school and would only talk to the "man of the house," she thought to herself. In this regard, the college diploma holding–Rosalie was just as feisty as Jacque.

She relented and got Randy. She asked him to take the call. Her twenty-four-year-old husband did. The conversation between the priest and Randy didn't last long.

Randy began to tear up. He hung up the phone, walked over and hugged his wife. "I am so sorry, Rosalie. Your sister Jacque is dead. It's going to be on the news shortly," said Randy to his wife, who was now crying out loud. Randy knew the pain of loss, as his father had died suddenly eight years earlier on the very property where he was now breaking the news about Jacque.

"Do my parents know?" Rosalie managed to mutter, clearly focusing only on the words *Jacque* and *dead* from Randy's first words to her after hanging up the phone.

"I don't think so," said Randy.

The couple placed a call to the Reedsville parish priest, who couldn't make it over to the Pritzl farm.

That's when the couple—married less than five years—made a call back to their Brillion parish priest, Father Dowling. He knew Ken and Jacque well, as they, too, were members of Brillion's St. Mary's Parish.

"I'll take care of it," Father Dowling assured them. "Stay by the phone. I'll call you back."

THE PHONE STARTED RINGING

"Hello, Pritzls," said Julia to the caller who had just rung her phone.

"Is there anything I can do to help?" asked the woman. Julia knew the caller by her voice as nearby neighbor Leona Burich.

"Not right now. Why would we need help tonight?" Julia asked.

"Oh, no reason," Leona said and abruptly hung up. Julia tried to call back, and there was no answer.

That's odd, she said to herself. Julia hustled out to the barn where her husband, Elmer, and Steve Reis were milking cows. Julia had a rare night off. She told Elmer about the call. He agreed it was strange.

"Stay by the phone and turn on the radio. Something is going on," the ever-astute Elmer advised.

She went back in the house, and Elmer and Steve kept milking cows.

Julia received another call: "I'm so sorry. Is there anything I can do?"

Julia recognized this caller, too. It was Elmer's sister Dottie—the same Dottie Elmer helped put through school years earlier.

"Why do we need help?" asked a growingly frustrated Julia.

Dottie, who was a registered nurse, clammed up at first. Julia pressed on, as Dottie and Elmer had a deep mutual respect for one another.

Jacque Konop is pictured here in the family farmhouse with her latest snowmobile trophy. One week later, her mother, Julia, would be answering the phone in the exact location as family and neighbors offered "to help." However, Julia still had not learned that her daughter Jacque had died. *Author's collection.*

Then Dottie said, "There was a news story about Jacque on the television." Dottie wouldn't give any other details.

With the 6:00 p.m. news broadcast over, Julia called the Konop residence.

"Hello, Konops," said a young girl. Julia instantly knew it was her granddaughter Julie. The two were quite close, and it was no coincidence that Julia and Julie were essentially the same name.

"Hi Julie, what's going on?" Julia asked her granddaughter in the calmest voice a grandmother could muster.

"Something has happened to Momma!"

Julia later wrote in her journal, "Then I knew it was 'all over' for my Jacque."

A Car Pulled into the Driveway

As Julia finished the call with her granddaughter Julie, she was looking out the window. Her daughter Rosalie's car had pulled into the driveway and pulled up by the milk house—not the farmhouse. That was highly unusual on a number of accounts.

For starters, Rosalie should have been home milking cows with her husband, Randy, Julia thought to herself as Rosalie stepped out of the driver's side. As that took place, the Reedsville parish priest, Reverend Neuser, appeared from the passenger side.

They went straight into the barn.

To this day, Rosalie barely remembers even traveling to the farm. Only her mother's journal jostled her memory years later, as Rosalie, too, was in a state of shock.

Father Neuser took the lead and walked into the barn in his full black priest attire and distinctive white collar. He had just celebrated the Saturday night vigil Mass.

"Elmer saw the priest walk into the barn, and he fell to his knees," said Steve Reis, who had never known Elmer to show much emotion during his tenure with the Pritzls.

"When Rosalie cleared the threshold, Elmer dropped the milking unit and started crying out loud. He knew full well that his daughter Rosalie should have been milking her cows with her husband, Randy, that Saturday night," continued Steve, who was a teenager at the time.

With that, Father Neuser confirmed Elmer's worst fears. His oldest child, Jacqueline Konop, was dead.

While this photo was not taken on that fateful day when Jacque died, this would have been an interior barn view of where Elmer fell to his knees on seeing a priest and his daughter Rosalie walk into his barn on that Saturday night with the dreadful news about Jacque. *Author's collection.*

Elmer's mind instantly flashed back to September 14, 1932. That was the day of his worst real-world nightmare—up until this moment.

On September 14, 1932, Elmer took a moment's break from his paper route delivering the *Green Bay Press Gazette* to investigate the commotion at the Brillion train depot. A woman had been killed by a train, and the townsfolk were trying to identify the mangled body. After jumping up on a bench, Elmer exclaimed, "That's my mom!" instantly recognizing the clothing.

The news Elmer was just learning was becoming the worst migraine he ever encountered. Not only had his daughter Jacque just died, but her two little children were about to grow up without a Mom as well. Elmer had lived out that painful experience, too. He knew full well that no human can replace one's mother.

As Elmer's mind was racing through these mind-numbing notions, Julia walked into the barn minutes later. She had the fateful notion that she had gleaned in the three phone calls—Leona, Dottie and Julie—confirmed by the priest, her daughter Rosalie and her husband in a never-before-seen prostrate position on the barn aisle.

Julia wept profusely.

By the time Elmer and Julia learned of their daughter's fate, media outlets had already beamed out the information on airwaves, and tragically, some next of kin still had not been notified of Jacque's death just hours earlier.

26

THE NIGHTLY NEWS TOLD OF
HER SISTER'S DEATH

Local news stations had created one big mess. Due to cases like the one involving the February 9, 1974 death of Jacque Konop, names of the deceased are no longer released until *all* of the immediate family has been notified.

While Rosalie and her big brother, Elmer John, had received word of their sister's demise, younger sister Annie wasn't in the know about the fatal accident that took place earlier that day at the Calumet County Fairgrounds. Annie heard from a different source—her mother-in-law.

The phone in her home rang.

"What did you think of the news?" said Mrs. Krueger with a nervous laugh.

"About what?" inquired Annie.

"Oh. Oh…let me talk to Robert?" replied Annie's mother-in-law.

Bob took the phone and walked in the other room.

Bob's mother proceeded to tell her son what she just heard on television.

"Breaking news, Jacque Konop, age thirty-three, was killed at the Calumet County Fairgrounds during a snowmobile race earlier," bellowed out the television reporter.

It was a short call between Bob and his mother.

He walked back in the room, crying.

"Somebody died." That was all Bob could muster.

Annie started guessing.

"Mom?"

Bob shook his head no.

"Dad?"

Another negative nod.

"Then who?"

Bob eventually gathered his composure to muster, "Your sister Jacque."

Bob hugged his pregnant wife.

To this day, it's the harshest news Annie has ever received. To this day, she never got over how she heard the news, either.

Several Green Bay area news stations earned an "F" grade in compassion and couth in airing that breaking news prior to family having a chance to learn it first.

These days, police and communication outlets have cleaned up their act, for the most part, so families learn about a loved one's fate before the rest of the world knows of their demise.

SORTING THROUGH THE AFTERMATH

What does a mother go through when a child dies? Julia's journal entries give a firsthand look into the grieving process.

"Jacque died February 9, 1974, at Chilton hospital," wrote Julia. "<u>None of us were at her side</u>," she underlined in an entry made prior to Jacque's funeral. "Just terrible!!!" she continued.

"Ken called the Brillion priest to come out and tell us. He called our priest, and Rosa and Butch were supposed to come and tell us. Finally Rose came with Rev. Neuser," wrote a frustrated Julia.

"Jacque had a 'full' church of people," she wrote after the February 12, 1974 funeral. "Pretty much all of Brillion and Reedsville showed up and students were crying throughout the visitation and the service," continued Julia, who made this notation knowing her daughter taught at Brillion High School and her son-in-law Ken was principal at Reedsville High. "The sun was shining nice at the cemetery. It was a warm day."

By the following Friday, February 15, Julia opened up far more in her journal.

"Well today is Friday—our beloved daughter Jacqueline died last Saturday, February 9, 1974, racing a snowmobile at the Chilton race," wrote the fifty-six-year-old mother. "She skidded on an ice spot (some said), flipped over, and the snowmobile flipped on her. Her neck was broken, and her lungs collapsed, too. She died about one-half hour after it happened, and a priest attended to her at the hospital.

Left: Taken weeks before her death, Jacque was juggling a full plate in the early 1970s as wife, mother, schoolteacher, advisor to four student groups and snowmobile racer. Ken and Jacque Konop are shown looking at some of their racing trophies with daughter Julie and son Chad. *Author's collection.*

Below: After their daughter Jacque's passing, Elmer and Julia set about showering their grandchildren Chad and Julie with love. *Author's collection.*

"She had a beautiful funeral and church service. Flowers—oh so <u>many flowers</u>. I picked out her clothes for burial," she continued, underlining *many flowers*.

"Mayme stayed with Ken for a week," she wrote of Ken's mother. "I don't know what they'll do. I can't do it—but I'll help with something," she wrote of Ken's future along with that of her grandchildren Julie and Chad.

"She was only 33 years old. But she wanted to accomplish a 'lifetime work' in the years that she lived.

"God take care of her. She's not dead, just away! May we meet in heaven!" Julia days later added "Someday!" in another tone of ink that clearly indicated Julia wasn't ready to make the trip to heaven just yet.

One year later, Julia made one more entry on the subject.

"It's February 9, 1975. Today, we had a mass said for Jacque. A year has gone by, and we've shed our tears," said Julia.

"Now we have to concentrate on other things, too; especially Julie and Chad—advise them as much as possible and shower them with care and love."

MUTUAL RESPECT

It's interesting what people will save. Tucked away in Julia's journal was a notecard. The front of it read, "The family of Jacqueline Konop acknowledges with grateful appreciation your kind expression of sympathy."

Carefully opening the card reveals its contents in impeccable cursive handwriting:

> *Elmer & Julia*
> *Thank you for the flowers and moral support you provided during our hardest days. Your presence and commitment to our family give me the strength to go on.*
> *Thank you and our prayers are also with you.*
> *Ken Konop and family.*

That short note struck a chord thirty-seven years later.

On October 1, 2011, Ken and his nephew, the author of this book, were having a brandy old fashioned mixed drink early on a Saturday afternoon at Brillion's Cobblestone Creek. The funeral mass for Julia had just been completed, and Ken wanted to talk. He chose yours truly for the conversation.

With a raised drink in hand, Ken turned and said, "To the best mother-and father-in-law a man could have in any lifetime. They could have tossed me off the property for getting their daughter involved in snowmobile racing. Instead, Julia and Elmer showered Julie, Chad and I with love."

The drink instantly disappeared, and tears soon flowed.

A FINAL TRIBUTE

"Synonymous Her Life...and Death," read the column headline in the February 12, 1974 edition of the *Green Bay Press-Gazette*. The column was written by John Lee, the newspaper's outdoor writer, just four days after Jacque died.

"Snowmobiling was a major part of Jacqueline Konop's life and family.

"Until Saturday, when the sport claimed her life in a race at the Calumet County Fairgrounds in Chilton.

"The 33-year-old woman, a teacher at Brillion High School and mother of two children, had been racing snowmobiles for four years and was considered a top competitor.

"Last season she qualified for the World Series of snowmobiling. This year, she ranked eighth in the Women's Mod One class of the USSA Midwest Division and would again have qualified for the series.

"Her husband, Ken, is also a snowmobile racer and is chairman of the USSA Drivers Committee. The two were members of the Green Bay-based Pabst racing team," he wrote. In those days, Pabst Brewery was a major sponsor even though Pabst had yet to reach its beer-making zenith. That would come in 1979, when it brewed a record 15.6 million barrels. The Konops had reached the big time.

"She enjoyed snowmobiling and along with Ken, she was dedicated to the sport, her family, and to life itself,'" team member Larry Jorgenson said in an interview with newspaperman Lee.

"As a racer," Jorgenson said, "she was not a hard charger but a steady, consistent driver. Off the track, she aided her husband in his duties with the drivers committee.

"On the track, Mrs. Konop was not just a racer but also handled some mechanical duties. She was petite and feminine, but she also did a lot of her own mechanical work. She was very knowledgeable in the operation of a sled in addition to knowing how to drive it," Jorgenson said.

27

SNOWMOBILE SAFETY HAS COME A LONG WAY

When Jacqueline Konop died on February 9, 1974, she was Wisconsin's twenty-first snowmobile fatality of the season. The very next day, Kevin Kubsch of nearby Kellnersville lost his life when his snow machine struck a guard post near Menchalville.

For every death, there were at least ten more reported injuries. That's according to the winter season data for 1972 to 1973 gathered by the Wisconsin Department of Natural Resources. Snowmobiling was gaining momentum as a recreational and racing sport, but safety gear hadn't caught up yet.

"People have long heard and read about or have been victims of traffic crashes," wrote Emmert Dose in Racine's *Journal Times.* "The National Safety Council reported figures it has collected on snowmobiling indicate a casualty rate about six times as high as automobiles," he noted in the February 11, 1974 edition of the paper.

Outdoor writers across Wisconsin were clearly shook up about Jacque's death and had made note of it, too.

A deeper dive into the Wisconsin Department of Natural Resources data for the 1972–73 snowmobiling season revealed:

31 deaths, 3 more than the previous season

342 reported injuries

Of the 679 reported accidents, 359 were due to the fault of the operator

114 of the 679 accidents were caused by excessive speed.

Dose also cited a Michigan State University study that found in 58 percent of accidents, drivers had violated traffic laws, while another 22 percent had consumed alcoholic beverages prior to the accident. Additionally, half of the fatal accidents occurred when the driver attempted to cross a public road or enter a road from a driveway.

Then there's the matter of rookies and daredevils. That's where the most carnage took place.

The most frequently injured are teenage and middle-aged men. The average age of snowmobilers involved in accidents was twenty-eight. Of all the accident victims, 18 percent were under the age of sixteen.

HUNKERED DOWN IN HIS SHOP

Looking to protect fellow snowmobilers after his wife's tragic accident, Ken Konop went to work. By December 1974—less than ten months after Jacque's death—he had engineered the Saf-Jac. Yes, it was named in honor of his late wife, with the "Jac" in Saf-Jac standing Jacque.

For Ken Konop, life would never be the same after Jacque died near instantly in a snowmobile racing accident. In honor of his wife's memory, Ken hunkered down in his shop and invented the Saf-Jac, named in his wife's memory. The protective jacket is still sold to racers. *Author's collection.*

Ken's jacket was made of a hard plastic and dense closed-cell foam. It maximized both protection and mobility. The jacket could stop penetration from racing hazards such as studs, skis and other objects.

It also dispersed the impact of those objects, such as a sled falling on a driver's torso, as was the case with Jacque.

Did it work?

One of the first users of the Saf-Jac vest fell off his machine in a drag race and was hit by another racer and thrown some thirty feet. The driver was unharmed.

"He had no scratches, no breaks, nothing—he just got up and walked away," said Ken. "We know it works."

In making the Saf-Jac, Ken also sought the help of doctors and chiropractors. The chiropractor bit was important, because many snowmobiler injuries involved the back.

He also had race drivers test it out. As a racer himself, Ken knew that the jacket could not impede a racer's ability to race. If the jacket was an impediment, they would not wear it.

When Ken finally went to commercial production, he settled on his seventh model. It retailed for forty-five dollars in 1974. By this time, Ken had also moved on from serving as principal at Reedsville High School, as he needed a job with more predictable hours and fewer after-school meetings. He was a single father with two little children.

OVER THIRTEEN THOUSAND SOLD

Saf-Jac did its job. It worked so well that Ken Konop became enshrined in the racing hall of fame.

"For more than 30 years, snowmobile racers and trail riders have enjoyed greater personal safety thanks to Ken Konop, inventor of the Saf-Jac protective vest and safety advocate," reads the citation in the International Snowmobile Racing Hall of Fame.

"Combining an impact-dispersing plastic outer shield with padded fabric underneath, the Saf-Jac was the first snowmobile safety product of its kind that protected the racer's torso. Introduced in 1975 during the first heyday of snowmobile race participation, the Saf-Jac has prevented countless serious injuries to thousands of thankful racers and riders, with an exemplary legacy of quality and safety."

Ken's induction to the Snowmobile Hall of Fame wasn't all about safety. Ken could flat-out race, too.

"Konop's racing career spanned 1966 to 1978, mostly in USSA's Central Division, having qualified for the World Series of Ovals for eight consecutive years, including a 1971 win in Booneville, New York. But it was the accidental death of his wife, Jacque, at a snowmobile race in 1974 that inspired Konop's greatest contribution to snowmobile racing history.

"Just months after Jacque's death, Konop developed, marketed, and sold the first chest- and back-protecting safety vests. Named in honor of Jacque, the Saf-Jacs set a new standard for personal safety and helped shape safe equipment requirements for all racers. More than 13,000 Saf-Jac protection vests have been sold throughout the world, averting untold injury to as many racers and riders for more than three decades. After retiring from 30 years as a high school teacher, Konop of Brillion, Wisconsin, continues to lend his expertise and help as an official in the USSA circuit."

Started Because of a Student

Ken got intrigued by snowmobiling in 1967 when Mike Miller, one of his students at Reedsville High School, told him about a snowmobile. At first, he had no idea what Miller was speaking about, so he went out to look at it and took it for a ride.

Ken was hooked.

In 1969, he used a sled from Potter Sports and won a few races. The following year, he qualified for the World Series in Boonville, New York. He won his division and was crowned a champion.

His was the only stock Polaris that qualified for the race.

"No one could believe a Polaris won the event," recalled Ken. "After the race, the tear down was in the horse barn, and many people crowded around to witness that it was all legal," Ken said with a smirk of the mandatory deep-dive into all the snowmobile's components prior to making the win official.

At the time of Jacque's tragic death, the couple had been racing for the Pabst Blue Ribbon Race Team. Jacque had won two series titles, and the Konops were only one of two couples on the racing circuit.

That was until Jacque died on February 9, 1974, in a snowmobile race.

Now other drivers can be protected by the Saf-Jac that bears Jacque's name to this very day. The teaching spirit in Jacque would be pleased that others are learning to be safer because of her death.

28

WHEN WISCONSIN WAS THE WILD, WILD WEST

Be sure of your target and beyond!'"
When it comes to hunting, that line is among the four tenets of firearm safety. One hunter's failure to follow that bedrock principle instantly dropped Kenneth "Ken" Kalies to the ground on October 10, 1974.

He was just forty-two years old and in the prime of his life.

Ken Kalies was born on December 12, 1931, to first-time parents Herbert and Mary Kalies, who lived in rural Cato back in those days. Known to his friends simply as Herb, he married Mary Burich on October 28, 1930, at St. Mary's Catholic Church in Reedsville. When the couple's firstborn son entered the world, Mary's younger sister Julia Burich became an aunt even though she had only become a teenager a few months earlier.

That also meant the young Kalies family became routine guests to the Burich Farm during Julia's high school years as they visited young Ken's grandparents John and Anna Burich. An avid hunter, Herb would also run his coonhounds in the Burich family's woods.

Ken eventually picked up his dad's love of hunting and fondness for coonhounds. He made trips to the farm as an adult, as the family homestead was the residence of his Grandma Anna until the day she died in the spring of 1951.

An avid notetaker, Julia surprisingly wrote very little in her journal about the sudden death of her nephew Ken. That may be because her firstborn daughter, Jacque, died earlier that year. Julia was emotionally spent.

"Ken Kalies got shot by a careless hunter in Wall's Woods near his home. He died October 10, 1974, of complications due to the gunshot wound as the bullet ended up by his spine," Ken's aunt Julia wrote in her journal, underlining the word *spine*. "Mary Kalies [Ken's mother] died September 8, 1972. Luckily, she didn't live to see it—a favorite Son!" penned Julia, ending her comments on the matter. Throughout Ken's life, Julia knew her nephew very well, as he had been a miniature bridegroom in her wedding in 1938.

To be fair, Julia's sister Mary loved all her children. However, Ken was Mary's firstborn of seven children, and he certainly held a special place in her heart—at least in Julia's mind.

Shown in his 1950 Wrightstown High School graduation picture, Ken Kalies joined the U.S. Army, rising to the rank of corporal. It was a careless hunter, however, not miliary service, that would take Ken's life. *Author's collection.*

Hunting Wasn't That Safe

"Trigger happy" might not even be an accurate portrayal of the 1914 Wisconsin hunting season. That deer hunt might have been closer to the Wild West.

Of the 155,000 deer hunters carrying a license that year, 24 were killed and another 26 were injured. On a percentage basis, that was Wisconsin's deadliest hunting season.

Safety concerns were a stumbling block for decades.

In 1970, deer hunters purchased 501,799 licenses and registered 72,844 deer that year. In addition, 13 hunters lost their lives in that hunt.

Lawmakers began understanding that hunting accidents had become an issue that needed a remedy. That's why Badger State politicians and agency officials initiated a hunter safety program in 1967. Early efforts ran with mixed opinions, as the program wasn't mandatory.

Deer hunting kept gaining in popularity. By 1978, over 644,000 Wisconsin residents went to the woods in search of deer. Recognizing that this number ranked larger than the entire assembled armies of many European nations, officials stepped up safety measures by requiring every hunter to wear blaze orange clothing in 1980.

Not satisfied with the safety metrics, the fledgling hunter safety program first initiated in 1967 received a promotion. By 1985, state law required all hunters born after 1973 to pass a hunting safety course to purchase a hunting license of any kind.

The Key to Safety

That mandatory hunter education was the right prescription. The statistics speak for themselves.

Just one decade prior to the mandatory program, the ten-year average firearms incident rate was 30 per 100,000 licensed hunters, according to the Wisconsin Department of Natural Resources. A decade after hunter safety became mandatory, that fell to 22 per 100,000 hunters.

Fast-forward to 2016, the incident rate plummeted to a miniscule 3.6 per 100,000. These days, you are more likely to be injured while bicycling than hunting.

That incredible safety standard has become possible because some 24,000 students become certified to hunt each year. According to the Wisconsin Hunter Education, ten to nineteen hours of instruction include wildlife ecology, hunting tactics and regulations, tree stand safety and, of course, firearm safety.

The Big Four

Safety—it's what makes a fun and memorable hunt. For anyone who has ever graduated from the Wisconsin hunter safety program, you learn T-A-B-K. Most hunters can recite the lines from memory. They are drilled into your psyche as a cadet in boot camp:

Treat every firearm as if it is loaded.
Always point the muzzle in a safe direction.
Be sure of your target and beyond.
Keep your finger outside the trigger guard until ready to shoot.

On October 10, 1974, one hunter in Wall's Woods did not adhere to T-A-B-K.

Ultimately, it wasn't a foreign enemy in Korea who killed Corporal Ken Kalies, who served in the U.S. Army from 1952 to 1954. It was a cavalier hunter near his Wrightstown home who ended his life.

On that day, Mary Ellen "Mickey" Kalies became a widow and six children—Becky, Maria, John, Terry, Tony and Alan—lost a father.

Months later, Mickey would place an advertisement in the local papers, "2 Palomino—Ponies. Harnesses & saddle. 2 coonhounds, Walkers, 10 months old."

With that sale bill, Ken's wife, Mickey, sadly sold some of Ken's favorite animals, ending the hunting tradition that Ken learned from his father, Herb.

29

LET THEM LEARN ON THE FARM

F arming may not be the highest compensated profession, but the children we raise and the integrity we instill in the next generation is worth more than any paycheck."

In the October 25, 2017 edition of *Hoard's Dairyman*, I captured that quote in the Editorial Comment page. No attribution was given.

It now receives its attribution: Elmer Wilfred Pritzl.

He said it when speaking about the interactions with the young men who worked on his farm over the years. Next to his wife and children, the mentorship of these young employees was his greatest source of pride.

One of those young men was Steve Reis. It's apparent that the feeling was mutual.

"It was amazing, those three years," said Steve.

"That was my best job…learning animal husbandry, woods, carpentry, mechanics, driving machinery and even operating a chain saw," Steve continued, recalling his teenage years working with Elmer and Julia on their Pine Haven Farms. "Elm trees were dying from Dutch elm disease, and we were cutting them down all the time and burning branches. Elmer eventually allowed me to run his Homelite chainsaw," said Steve. "That was a big deal for a high school student in that era.

"When I started my painting company, I pulled directly from my knowledge of wood. By the time I started working for the Burger Boat Company, I thoroughly knew how to identify wood species by just looking

For Steve Reis and a host of other high schoolers, learning to operate heavy equipment and work with farm animals truly brought education to life. *Author's collection.*

at the grain," continued the 1970s-era Reedsville High School student who now owns the Woodfire Lodge at Triple J Wing & Clay, among some other business endeavors.

LEARNED TO WORK

"I thought I knew how to work. Then…well, then I met Elmer," said Steve.

"One day Elmer had to take Julia to the hospital," he said, noting that he later learned Julia had a hysterectomy and Elmer had to be away all day.

"Dig holes for a new fence line," Elmer told Steve. "When I get home, I will set the new fence posts so that the cows can graze this field."

"Elmer then gave precise detail on how to go about measuring the fence hole distance and maintaining a straight line," said Steve, who was a high school student at the time in the early 1970s.

"Back then there weren't automatic post hole diggers," recalled Steve. "So, I started digging holes with a post hole digger…one hole at a time, all by hand. I had a good day and dug over thirty holes. I thought, wow, this is going to take Elmer days to set posts in those three-and-a-half-foot-deep holes and repack the dirt.

"Not so," said Steve.

"I came out to the farm the next day after school and every one of those holes had a new post in it with the fence line already attached to it.

"Can you dig some more holes?" Elmer asked of Steve. "We might get this fence done by week's end."

"I was astonished. A man twice my age was somehow outworking me," said Steve. "Then I learned Elmer's trick.

"Elmer always took a nap at noon, after already putting in a seven-hour workday. He would get up and be ready for another eight hours," said Steve, who worked on the farm from 1973 to 1975. "Amazing."

That was just one of Steve's many memories from the Pritzls' Pine Haven Farm. Some additional recollections include:

MECHANICS: Elmer had a self-propelled Owatonna hay mower in the 1970s. "Learning to drive a machine with only brakes and stick steering prepared me for driving anything," Steve said.

DRIVE STRAIGHT: He wanted every row to be straight, whether it was in a field of hay or corn. "Put your eye on the horizon, look for post or tree and aim for it. Once that first row is straight, so will the rest," advised Elmer.

WOOD MAKING: Elmer loved trees. "We would cut trees from his 'Pine Woods' across the road from the farm and skid them out with his WD Allis Chalmer tractor," Steve described. "By making a ramp with two trees, we would skid those logs on the wagon and haul them straight to the Schleis sawmill in nearby Menchalville. We'd often stay at the mill and help cut logs. It was a learning center as I watched pine, maple, white oak, red oak, black cherry and even black walnut come off the mill. Elmer explained the grain structure, and I learned to identify wood. He could have been a college professor when it came to his ability to work with wood and teach others about trees."

He continued, "Later, we would stack that lumber with slats between it so it could dry. Then I could study the grains even more."

The reason Elmer rolled his own cigarettes dates back to his days working at the Brillion Iron Works and its foundry. It turns out it was so hot and sweaty in that building that regular packs of cigarettes got rather soggy. *Author's collection.*

"We also walked though his woods and planted black walnuts everywhere. He'd use a metal pry bar and poke a hole in the ground. I'd follow with the nut. The next spring, many of those nuts sprouted into trees."

CARPENTRY: "He was an engineer. I watched him take a flat metal spade and file it down. He made a perfect pattern that allowed him to make well over two hundred spice cabinets. That was something to see. And if he made something, it was 'Elmerized.' By that, I mean he built it to last."

PERSISTENCE: "Watching him fix things was amazing. He generally figured things out and seldom called for help. He would use WD-40 and slowly loosen up radiators that most people would have tossed away. Eventually, after weeks of applying WD-40, we'd get them apart and repair the fins. If something wasn't going well on a repair, he'd walk away, sleep on it and generally develop a workable solution," admired Steve.

LEARNED TO LISTEN: When giving directions, Elmer had some German tendencies acquired directly from his first-generation American parents born of immigrants from Deutschland. That being the case, he expected and received full attention when giving directions for a project. And the penalty was simple for not listening. "He would not repeat the directions, and you would have to figure it out yourself," Steve remembered.

"I learned to listen to detail and listen the first time," added Elmer's daughter Annie. "Dad wouldn't tell you a second time. You were definitely on your own."

THE WHIMSICAL MEMORY: Those who knew Elmer knew he liked to roll his own cigarettes with Top Tobacco paper and Velvet Pipe Tobacco. No filters, of course. "While I wouldn't recommend it, imagine watching a grown man rolling his own cigarette with one hand while driving a tractor," Steve said. "He was that good with his hands."

THE TRADITION CONTINUED

When Randy and Rosalie took over the farm, they continued Elmer and Julia's tradition with over a dozen high schoolers eventually working on the farm. One of those young men was Tad Campana.

"Do you think they can work?"

That's the question that Tad now asks himself every time he reviews a résumé.

"Will they do what it takes to get a job done?"

"From working on the farm from 1985 to 1991, I learned to do that firsthand, and because of that experience, I can spot it on a résumé or very early on when talking to a potential job candidate.

"My first job on the farm was indeed a classroom experience," said Campana, who learned from both Elmer and his son-in-law Randy, "but in this case the classroom had no walls, as it was an outdoor classroom on the farm."

30

YOU STOLE MY TREE

In the spring of 1986, I went for a walk in the Pine Woods with a shovel, a five-gallon pail and two escorts: Snoopy and Buttons, the farm dogs. My mission on that Sunday afternoon was to get one Wisconsin white pine and transplant it on the farm. I was in the sixth grade.

Unbeknownst to me, Grandpa and Grandma decided to go for a drive and take a spring walk in their woods. Because my grandparents were forever bringing food scraps out to the farm for the two white terrors, those dogs never barked out their arrival. Suddenly I heard, "What are you doing with that tree?"

It was Grandma. At first, I thought Grandma was joking. Then I thought to myself, "Oh boy, she's serious and I'm in trouble."

At that moment in time, Grandpa and Grandma Pritzl still legally owned the forty-two-and-a-half-acre Pine Woods, as it wasn't a part of the 1981 purchase by Randy and Rosalie Geiger. It just felt like "ours" to me, as it was my outdoor classroom where Grandpa Elmer, Dad and I made firewood every Saturday throughout the winter and where I learned to identify every native Wisconsin tree without the aid of leaves, courtesy of Professor Pritzl.

Years later, the Pine Woods would be purchased and transferred in a unique fashion with 50 percent being held by my mother, 25 percent by my sister and 25 percent by yours truly. It was the only piece of the farm sold that way, as my grandparents wanted to ensure one heck of a debate would ensue if the word *sell* was spoken aloud. The patriarch and matriarch clearly did not want that property divided or sold out of the family.

While taken later in the summer of 1986, author Corey Geiger was beginning to realize the learning opportunities before him in Grandpa and Grandma Pritzl's outdoor classroom. *Author's collection.*

On that day, I was served a big piece of humble pie. To ensure I knew what Grandma had on the lunch menu, Grandpa quietly took two steps behind Grandma and gave me a look that could only be interpreted one way, "Please be smart. Grandma must win."

The telepathic communication worked.

"Grandma, may I have one little pine tree? Please," with the *please* added at the end because the stern look was still on Grandma's face. "I can put the tree back where I found it if you like. However, I dug it out from the center swamp. If it goes back, it will tip over like all the others and die. That would be sad for the tree and for me."

My plea and groveling seemed to be working. A small smile appeared on Grandma's face. "Sure, I guess you can have the tree. But you better ask in advance next time if you have plans to do this again," she added as a measure that I better respect my elders. Grandpa, still standing behind Grandma, winked at me to indicate a job well done on that spring 1986 day.

31

BOARDS CAN TELL STORIES

As teenagers, we think we know everything. I was no different at that age. To be honest, looking back I had just begun to understand the world around me.

Such was the case in the summer of 1992, when I turned nineteen. That year, we completely remodeled our dairy farm's milk house. The job included installing a new bulk tank and compressors for storing and cooling milk. As part of the process, we tore down the old walls, constructed new walls and added improved insulation to keep the facility warmer in winter.

With the help of Bob Behnke, who lived nearby on a farm nestled between Long Lake and Becker Lakes, I, quite frankly, thought we had done a bang-up job. That was until Grandpa Elmer came upon the scene.

"Where did you put those boards after you tore down the milk house walls?" asked my grandfather in an unusually firm voice.

Unsure of the direction of the question, I simply said, "I tossed them in the farm gravel pit. They were junk."

"You didn't!" exclaimed Grandpa Elmer.

It was at this point, I knew he was serious, so I started backpaddling.

"If you really want them, I can help pull them out," I sheepishly said, knowing it was better for a nineteen-year-old to go dig for them than watch my seventy-two-year-old grandfather venture in after them.

Off we went. I should have known better. As we traveled to the gravel pit, my mind was racing, and I recalled a carpentry project from a decade earlier. Back then, Grandpa Elmer taught me another lesson after he caught me pounding nails erratically with my hammer to secure floorboards in the haymow.

"Why are you putting in all those hammer heads into my pristine pine boards?" he sternly asked.

"No one will see them," I quipped back to him.

"You get an F! You must learn to pound nails without making hammer heads. This is perfect practice for more important projects. Now do better!" he said, walking away. For the rest of the day, I felt like one of Grandpa's eyeballs was riveted to me, observing my every swing. Of course, I did better, as an inspection did come later.

Upon arriving at the gravel pit, we pulled every last one of those boards out, old nails and all. It wasn't an easy undertaking, as each board was two to three feet wide. Yes, that was their actual width, and I was going to learn why really quick. As I pulled each board out of the deep pit, Grandpa Elmer carefully set it into his 1970 red Ford pickup truck, looking each one over for any damage.

THESE ARE HISTORIC

After the boards were secure, Grandpa sat on the tailgate, and I knew right away that it was story time.

"Do you know where those boards came from?" said the history lover as he rolled a cigarette using tobacco paper and poured tobacco from a can kept in his shirt pocket.

"No idea," I said.

"It's very likely that Julia's grandfather and great-grandfather nailed those boards onto the walls or roof of their first home back in 1867," he started out. "Those boards came from virgin white pine timber. That's why they are two and even three feet across," he explained. "Those boards were cut from trees that stood prior to European immigrants coming to the Great Lakes region. Those came from Wisconsin's great white pines."

I had no reason to doubt my grandfather, as he had a keen memory. After retiring from forty years as a dairy farmer, he took up carpentry as a second career.

"After building a new faced cement block home in 1916, my father-in-law, John Burich, tore down the pioneer house board by board…saving each piece," said Elmer puffing away on his cigarette. "The very next year, he built a new barn for heifers and horses and reused those circa 1860 boards as roofing material. He also used some to build the milk house."

Elmer knew all the fine details of his adopted family farm. Elmer (*left*) often employed his father, Louis "Louie" Pritzl Sr., to help with carpentry and masonry projects. Daughter Rosalie sits on the rock pile. *Author's collection.*

Sure enough, I had remembered seeing a number of new boards in the roofline while stacking hay in that barn during many of my summers on the farm. There was a section of that roof that had much darker, incredibly wide boards.

"Those old boards came from the homestead house?" I asked, half making a statement and half asking a question.

"Yes, they did. We reused everything back in those days. People are so quick to throw things away these days," commented the man who lived through the Dirty Thirties and the Great Depression.

"What will you do with these now that we pulled them out of the gravel pit?" I asked.

"They will have a third life as benches," said the carpenter who practiced green-minded sustainability strategies before the topic was in vogue.

And that is what he did, making four benches from the virgin white pine timber boards planed 125 years earlier.

BIG PINES WERE BIG MONEY

Where did Elmer Pritzl first gain an appreciation for the pine trees? His father, Louis "Louie" Pritzl, of course.

Back in the day, Louie would work winters in the logging camps in Wisconsin's Northwoods. Many immigrants and the ensuing generations' adult children would work in these lumber camps.

In fact, according to the 1890 census, more than 23,000 men worked in Wisconsin's logging industry, and another 32,000 worked at the sawmills that turned timber into boards. That's from data gleaned from the Wisconsin Historical Society. Each winter, lumberjacks occupied nearly 450 logging camps. In spring, those logs were driven downstream via rivers to more than 1,000 mills.

Louie Pritzl was an expert logger.

One winter, he and his axe-wielding lumberjack partner were asked to fell one of those big Wisconsin white pines. However, the logging boss had this offer, "If you get that pine down without breaking it, there is a can of

Louis "Louie" Pritzl Sr. was one of forty-one men in this Northwoods lumber camp. Wisconsin had over 450 such camps that employed twenty-three thousand men. Louie is shown in the front row, sixth from the left. *Author's collection.*

tobacco in it for each of you," said Elmer, recalling having heard the story many times before from his father. That pipe tobacco was a big bonus in those days considering the low wage scale.

The duo spent a good part of the day chopping away. And by day's end, they indeed felled that one-hundred-foot-tall tree measuring well over three and a half feet in diameter into one continuous log. For Louie, it was perhaps his fondest logging story. And yes, he enjoyed his tobacco, too, as did most lumberjacks. It's these stories that created the fabled legend of Paul Bunyan. And that's where Elmer began to learn that boards could indeed tell stories.

32

MAKE ME A DUCK

How will you spend your golden years? Hopefully, fulfilling your remaining bucket list of dreams.

For Elmer, that eventually included hand-crafting ducks. That flock of wood ducks were among dozens of projects that Elmer would produce in his sprawling woodshop. Of course, he built that shop, too.

By 1989, Elmer, age seventy-two, was eight years into his third full-time profession of carpenter, with his first two careers being a cupola foreman at the nearby Brillion Iron Works years earlier and later as a dairy farmer.

On the duck-making front, by no coincidence, 1989 was the year that Julia purchased a pair of books: *Making Wood Decoys* by Patrick Spielman and *An Artist's Portfolio of Wooden Ducks* by Dave Ladd. True to form, Julia, the diligent accountant, recorded the cost of each book on the title page. In the book *Making Wood Decoys*, she recorded $9.95, plus $2.20 shipping, and added the special note, "Make me a duck."

And so, Elmer went about creating a of flock of wood ducks that mirrored the flocks of real Mallard and Peking ducks he raised decades earlier on the farm. The wood ducks would come from nearly every tree species he harvested from his woodlots—except for elm and oak, as that wood was too hard to carve. Black walnut and butternut were his favorites.

Because he admired the beauty of the grain, he just added a set of eyeballs. The day each duck "took flight," he signed his name on the bottom *Elmer Pritzl* and inscribed their birth date, just like the records for the newborn calves on his farm years earlier. And just like their days milking cows together, Julia pitched right in to finish each duck with a coat of stain and varnish.

Real-life Mallard and Peking ducks Elmer raised years earlier would become the inspiration for his wood ducks. *Author's collection.*

The butternut duck adorning my writing desk was born on October 1, 1990, the beginning of my senior year of high school, and my black walnut duck took flight on September 15, 1994, just two weeks before I entered my final competition as a member of the University of Wisconsin–Madison Dairy Cattle Judging Team.

A SUSTAINABLE CARPENTER

Elmer had already become a master carpenter before he started making ducks. In the late 1940s, he chopped down enough Wisconsin white pines from his woods—by hand with an axe—to become the rafters in his soon-to-be-built one-hundred-foot-long by thirty-foot-wide machine shed.

He used small pine trees and turned them into four-by-four-inch-diameter mini beams with the center of each one being the original growth. That made them less prone to warping and thus stronger roof supports, he shared with me years later. Larger pines were cut for shed doors and the roof.

Because he wanted a smooth granary floor to push oats, he cut down hard maple trees and milled those into floorboards. To this day, it's the best granary floor my eyes have ever seen, and if that shed ever comes down, that floor could have a second career as a basketball court. Yes, it's that good, polished, smooth and ready for a team of cagers.

After that endeavor, Elmer salvaged the virgin Wisconsin white pine planks encasing the chimney in the pioneer log home of Julia's great-grandfather Wencel Satorie. He gave those planks a second life as a bench on which I sat for the author photo of my first book. Next, he refurbished Wencel's handmade chair that rocked Julia's mother, Anna, to sleep as a baby.

About that time, he cut the largest tree on the farm and hauled the forty-five-footer to the mill on a sleigh, shoveling snow across U.S. Highway 10 so the metal runners would better skid across the paved highway. Once the wood was cut into form and hauled back to the farm, Elmer assembled a team to hoist the beam into place to repair the upper story of his main barn housing the dairy herd. From there, he built a two-sided crib, which was a shed that could store cob corn on the outside with room for two tractors in the center.

Just like the machine shed a decade earlier, he cut all the pine trees for that, too, and quarried the gravel from the farm's nearby pit.

In one of his largest carpentry endeavors, Elmer cut and milled all the lumber for his one-hundred-foot-long by thirty-foot-wide machine shed. *Author's collection.*

Elmer continued to log trees and mill lumber well into his early eighties. *Author's collection.*

He regularly logged trees with his WD Allis-Chalmers tractor and hauled the logs to a nearby mill. After the lumber was milled, it was brought home and stored outside with wooden slats to allow it to air dry without the expense of a kiln. Sometime later, he would restack the pile, flipping the boards over for an even dry down. A few years later, the finer lumber such as the black cherry, black walnut and red oak would be placed back on a wagon and make a second trip to the mill to be planed.

Elmer lived sustainably, sourcing nearly all his building materials from the family farm. This was well before sustainability became an everyday word in our lexicon as attention has turned toward implementing strategies to mitigate climate change.

THE "BARN RAISING"

By the time 1981 rolled around, Elmer had gone about crafting the rafters from his white pine lumber inventory to build a massive two-story garage on his double lot in the village of Reedsville. Like in previous projects, all the wood was cut "old school." There were no one-and-a-half-inch by three-and-a-half-inch two by fours. These were the real deal, a full two-by-four-inch-width lumber.

Right before Elmer put his final plan into action, he explained the project to his daughter Annie. She suggested that a thirty-two-foot-long garage was not long enough. Upon overnight reflection, Elmer called his youngest daughter back and said, "You're right, I'm shifting the plans to forty feet."

After making a few more rafters, he literally held a barn raising—or in this case a garage raising—over a weekend, and the entire building was in place. Thrifty Elmer had only two vehicles, a car and a truck, to store in that massive building. The rest of that facility became his woodshop, and the second story was filled plumb to the rafters with his milled, home-grown lumber.

That's when he doubled down on his carpentry endeavor.

"Elmer restored antiques, crafted maple desks for all ten grandchildren, built two spinning wheels from scratch, mounted an assortment of deer antlers, and created benches, cabinets, nightstands, footstools, and spice boxes. Anything your heart desired," wrote Julia in her journal.

Those spice boxes were perhaps the most impressive, adorned with hand-carved oak leaves and acorns. Spice box No. 1 rolled out of Elmer's shop in 1988. To craft these pioneer-replica spice-storage units, he saved

Early in his carpentry career, Elmer made ten desks, one for each grandchild. *Author's collection.*

Elmer crafted over one hundred spice boxes, often adorned with hand-carved oak leaves and acorns. These boxes were made from maple, apple, red elm and black walnut, respectively. *C. Todd Garrett.*

every scrap of wood from larger carpentry projects. Of course, he crafted spice boxes from every tree species from his woodlots, including red elm and the oaks.

As for the very first spice box, Grandpa and Grandma asked what I wanted as a wedding present. And the day before our wedding on October 19, 2007, Spice Box No. 1 came to live with us.

Because of Elmer's skill, Ken Kuether, a local cabinet maker, would routinely hire him to help with in-home installations. Elmer didn't really do it for the money; he just wanted the challenge. One time Ken returned the favor and became Elmer's assistant when the duo installed a series of shelving in his daughter Annie's house. Fitted with an exquisite series of handmade ornate dowels to showcase plates, Elmer used the sickle bar from his old hay mower to give it perfect artistic appeal. Ken was impressed and borrowed the idea in later cabinet work.

WHERE HE GOT HIS START

After graduating from Brillion High School's class of 1934 as a sixteen-year-old, Elmer went straight to work for the Brillion Iron Works. At the tender age of eighteen, the young lad was promoted to cupola foreman. That role saw him supervising men twice his age as his team fed pig iron into the top of the cylindrical furnaces or cupolas for re-melting, along with small quantities of scrap iron, coke for fuel and limestone to act as a flux.

Five years later, Elmer and Julia entered into an agreement to run the family farm. This took place much to the chagrin of Butch Peters, long-time president of the Brillion Iron Works, who hired Elmer in just his second year of leading the Brillion Iron Works.

Butch had always hoped Elmer would go to college, become an engineer and return to work at the company. After one final plea at the family farm before World War II broke out, Butch gave up on that idea. But he was right about this—Elmer had an engineering mind and put it to work on his farm and later in his woodshop.

33

WILL THE FAMILY BUSINESS CARRY ON?

Their golden years were far more enjoyable because Elmer and Julia could rest easy. That's because they did their homework years earlier to best prepare their beloved family farm business for a successful transfer to the next generation. In doing so, they did their very best to prevent the four deadly Ds—debt, death, divorce and disability—that can conspire to take down a family farm or any business for that matter.

To be fair, no one can avoid death, but the other deadly Ds can arrive when we least expect. However, it's the plans put in place that help shield a business when one, two, three or the worst, the four deadly D grim reapers arrive at the business gate.

As a person who has covered agribusinesses across North America and around the world, I would add two more "Ds" to the deadly D list: "inDecision" and worse, "no Decision." Once a business begins to unravel, resuscitation attempts almost always fail. The business funeral then unfolds as assets are torn away. When it comes to family-owned enterprises, these funerals can be quite frequent, as 90 percent of America's businesses are held by families according to data from the U.S. Census Bureau.

"InDecision" and "no Decision" come into play when the senior generation never formalizes a plan to transfer the farm assets. Verbal promises are made to the junior generation but never fulfilled in writing and signed.

In so many instances, the verbal assurance is made to a hardworking son or daughter who remains on the farm, often working like a mule believing that he or she is earning sweat equity. However, the transfer agreement is

Transition planning and transition agreements can make or break a family farm. Gathered in 1967, Elmer and Julia stand front and center as the extended Burich family gathered to celebrate the family farm's centennial and four generations of successful farm transfers. *Author's collection.*

never recorded on paper, signed and notarized. As the years go on, other siblings develop careers of their own off the farm as adults. For the junior generation, life continues with the belief all will be well in the end and fairness will prevail.

THE BIG GRAB

Then the inevitable comes—death. And just like the famed story *'Twas the Night Before Christmas*, off-farm adult children begin to see visions of sugar plums dancing in their heads. The sugar plums in this case are the farm assets, mainly land that has greatly appreciated in value over the decades. And with no documented plan, everyone starts jockeying for their share of the sugar plums.

That's when the saddest affairs start to unfold. In some instances, the farm is busted apart because the junior generation cannot afford to buy out his or her fellow heirs. Conversations cease, and this tears the family farm fabric into shreds. In some of the worst scenarios, brothers and sisters never speak again.

Think it's not possible?

Sadly, I've seen it many times. The farmer in me cries internally every time the story unfolds. My pain is for both the family and the farm.

As for a fair way to allocate the farm assets?

For decades, I've advised that equal and equitable are not the same when it comes to farm businesses. Equitable, in this case, would be rewarding the farming child with a far greater share for working side by side with the senior generation. However, that is a personal decision for each farm family.

Most farmers need extra support to get this process rolling. Start with a trusted adviser—could be a banker, attorney, extension agent or another confidant. Then, start having conversations—lots of open communication. It is a must. Sometimes the parties will not like what they hear at first. The point is to keep the highway of contact free and clear. Then, go to work documenting those conversations with an attorney.

The First Said, "No"

While Julia had always wanted her son, Elmer John, to take over the family farm, the son bearing the name of both her husband and father had no interest in doing so. That was revealed in the first round of conversations in the late 1970s.

Julia pushed a bit harder, as she always dreamed that her son would take over the farm. But the more Julia pushed, the more her husband, Elmer, understood that his son enjoyed his profession as a machinist in nearby Two Rivers.

There also was another reality—Elmer John was completely done with the farming idea. He believed his father was a workaholic and that his childhood was buried on the farm. He often confided to his closest friends, "I didn't leave the farm, I escaped." In some ways, that assessment would be fair, as Elmer John worked on the farm before the family purchased silo unloaders and barn cleaners. That meant pitching feed and manure was a hand labor endeavor.

What was the senior generation's next plan?

That's when fate once again intervened. Elmer and Julia's daughter Rosalie and her husband, Randy, were very interested in becoming the third

Left: Elmer Pritzl realized that his son, Elmer John, was not interested in the farm. However, he was beginning to understand that his daughter Rosalie and her husband, Randy Geiger, had keen interest in becoming the fifth generation to run the enterprise. *Author's collection.*

Below: Over the course of four decades, Elmer and Julia Pritzl gradually grew and improved the family farm, shown in this 1978 image. *Author's collection.*

successive generation to have the farm pass through the family's daughters. The younger couple had been looking for a larger farm, and the Burich-Pritzl enterprise was about perfect. Randy had already been routinely helping Elmer with field work.

More discussions ensued.

Just like generations before, the equivalent of a land contract bound the final deal. No banker was needed, but attorneys were employed to knit the plan together in formal words that included a 9 percent rate. As for improvements such as a new silo, a pair of barn cleaners and two complete barn remodels, that involved a bank loan at the whopping rate of 18 percent. In the fall of 1980, Elmer sold his cows, and the fifth generation began remodeling the empty barns.

Then on April 4, 1981, the senior and junior generation went to Attorney John Stangel and signed the deal in nearby Manitowoc. That May, Rosalie and Randy's Holstein herd arrived, and by summer, Elmer and Julia had completed the renovations to their Reedsville retirement home and construction of the massive garage for Elmer's cabinetry work.

The entire land base was not transferred at first. Elmer and Julia held onto their beloved 42.5-acre Pine Woods along with a 2.2-acre lot on the opposite end of the farm in which they had hoped to eventually build a new retirement home years later. Due to open dialogue initiated by Randy and Rosalie over a decade later, those parcels were purchased by the junior generation to make the farm whole again.

As all of this was taking place, the genesis for this book began. Even though Elmer and Julia lived in town, their heart was on the farm, and they frequently traveled from Reedsville to the place they loved the most. And I became a sponge soaking in their stories.

Not All Plans Stay in Place

My parents began making monthly payments to Elmer and Julia. They were $1,869 per month, I recall. Then came a run of dry years, as 1985, 1986 and 1987 became successively drier. The summer of 1988, however, topped them all. It was the worst drought either couple had ever seen.

My father contemplated throwing in the towel on farming altogether. I still remember the day clearly as the would-be auctioneer walked up and down the barn aisle, assessing the herd for a potential sale. After the auctioneer left, I pleaded with all my heart outside our milk house that evening. It

On a warm spring day on May 17, 2003, Elmer and Julia Pritzl would gather at the family farm for their granddaughter Angela (Geiger) Zwald's wedding day. Pictured here are (*left to right*): Elmer John Pritzl, Rosalie Geiger (née Pritzl), Annie Krueger (née Pritzl) and Julia and Elmer Pritzl. The day reminded the family patriarchs of their wedding day sixty-five years earlier when family and friends also gathered at the family farm. *Jim Leuenberger*.

caused Dad to pause. Ultimately, he passed on the sale and soldiered on that summer, ultimately creating my destiny.

As for Grandpa and Grandma, they knew it was going to be tough. For these life-long farmers, they later said that one year was more desperate than any one year they witnessed during the Dirty Thirties. Elmer and Julia stepped in and suspended land contract payments. At first, payments were rolled to the end of the land contract. As the corn crop shriveled up later that summer, Elmer and Julia even forgave a few payments as an early inheritance and then paid out equal money to the nonfarm siblings.

Mom and Dad quickly diverted the extra money to buy irrigated alfalfa hay from out west to feed the dairy herd. We often unloaded hay late into the night that summer as the oppressive heat made it near unbearable to unload semis during the day.

Just how bad was it that year?

Elmer and Julia Pritzl's smiles were a bit brighter in their retirement years knowing the family farm was successfully transitioning to the next generation due to careful planning, written documents and continued conversations. The couple were forever frugal farmers—note Elmer's patched jeans courtesy of Julia's seamstress work. *Author's collection.*

Julia placed the July 4, 1988 issue of *Time* magazine into the family Bible. The cover read, "THE BIG DRY." She added in pen on the cover, "We lost money, too!!!" as to the family's plight that year.

I share all of this because many farmers never really retire from farming. Farmers keep working the land and with their animals because it's their identity; farming runs through their blood. And for Elmer and Julia, this was also true—their hearts lived at the farm.

The "never retire from farming" can work—if there is a written plan for the asset transfer. Those plans can be mutually amended during dire times, just like the big dry of 1988. However, without a plan for both the senior and junior partners, laying out the all-important income stream for all of the families, the family farm may vanish when deadly Ds begin dropping bombs on the family business.

Elmer and Julia had a plan. So did Rosalie and Randy, as they helped craft it, too. It involved a well-thought-out asset transfer, and that helped Elmer enter his third career in carpentry.

34

THE TREES COME HOME

All good things must come to an end.

Originally penned by the medieval poet Geoffrey Chaucer, who wrote the epic *Canterbury Tales*, that phrase began to apply to Elmer's vibrant career as a carpenter. On January 2, 2002, he signed an unfinished wood duck *Elmer Pritzl*. His penmanship lacked its customary stylish flow that earned him the letter grade of A from the Franciscan Sisters who taught him at St. Mary's Catholic School.

The letters wobbled.

And even worse, he kicked the duck out of the nest early, as it didn't get a coat of stain or varnish. Frustration clearly began to set in when the ducks no longer met his high standards. Elmer had overcome two dreadful bouts with lymphoma, a cancer of the lymph nodes, years earlier, but this was different. That's when he put plastic ribbons around the necks of the final four as if sending out a plea for help. Elmer's woodshop began to idle as his once steady hands and great eyesight both began to fail him. High schoolers who worked on the family farm began to take over snow shoveling and lawn mowing duties.

Dementia began eating away at Elmer's engineering mind and his impeccable memory. Macular degeneration also began to cause the loss in the center of his field of vision. It's about this time that Julia suggested that their subscription to *Hoard's Dairyman* be suspended, as he wasn't keeping up reading it. However, Grandpa Elmer still had some good moments.

Frustration set in when his ducks no longer met Elmer's standards. That's when he placed a ribbon around the unfinished ducks (*shown center*) as if sending out a plea for help. *C. Todd Garrett.*

A TWINKLE IN THEIR EYES

About the same time these health events were unfolding, my parents walked side by side with me to purchase an 80-acre parcel of land that had been owned by neighbors Ralph and Florence Moede since 1946. My parents co-signed the bank loan, as I had little in the way of collateral. My dad kept the land deed in his safety deposit box as his insurance policy just in case I didn't fulfill my end of the bargain. The interest rate was just over 8 percent.

Aside from the payments that strapped me financially, leaving only $200 in my monthly budget, I was excited to have my own piece of planet earth. That acquisition also grew the family farm from 215 to 295 acres and provided enough land base to more than feed our dairy herd.

That land purchase reinvigorated both Grandpa and Grandma, who were now both well into their eighties. Remember, once a farmer, always a farmer. I could see a little more bounce in Julia's step as her family farm was growing again. She could recite the ownership lineage of our new property straight back to immigrant Wisconsin, and it turns out Ralph and Grandma Julia had been classmates in the Reedsville High School class of 1936. Grandma and Grandpa were pleased.

And so, Elmer and Julia would drive out of town to investigate what their grandson had bought. Using his cane, Elmer slowly looked around the dilapidated barn and found some old pulleys and hay hooks. He had enough wherewithal to pluck those out of the barn and refurbish each one based on his childhood memory. That Christmas, those treasures were returned to me as presents.

At that time, the frugal couple was also frustrated with the interest rates they were receiving at the nearby bank on their certificates of deposit. That's when they walked into the bank to pay off my loan. Of course, the bank wouldn't let them do that. So, they withdrew part of their investments to cover my land loan, and on January 15, 2002, we entered a land contract of sorts. It was a double win, as I borrowed the money directly from my grandparents at half the interest that I was paying at the bank and Grandpa and Grandma were earning twice as much interest on their certificates of deposit.

The Quick Repayment

I hustled to pay off that note, as I could see the dramatic shifts in their health. Every month, I hand delivered the repayment checks. Sure, I could have placed those checks in the postal system. However, during those visits, I was inspired to ask the questions that in due course, breathed life into this book. It was during this repayment journey that I found the courage to ask about the day the train killed Elmer's mother, Anna.

The day I handed Grandpa the final check to repay the loan in February 2005, Grandma answered the front door and quietly gave me explicit instructions. "Please look Grandpa in the eye and tell him this is the last check. We'll need to sign a document, too, so I can show it to him whenever he asks. His memory is kaput," she said with a tear running down her face and having used the German-origin word *kaput* for extra emphasis. And so, I repeated the message several times that day. What should have been a happy

By the time Corey Geiger paid off the land contract to his grandparents, dementia had overtaken Elmer's once photographic memory. Despite Elmer's condition, family helped Julia and Elmer Pritzl attend the wedding of their grandson Corey and his bride Krista Knigge in October 2007. Elmer passed away nine months later. *Author's collection.*

moment to own land free and clear turned about to be a sad day of sorts. I saw emptiness in Grandpa's eyes.

Just a few months later, the next dreaded conversation was forced on us. Julia called the farm phone. Having just turned off the milk pump, Rosalie asked longtime herdsman Dan Dvorachek to answer the phone. "Danny, can someone come over right away? Elmer backed the car through the closed garage door. It's not good."

Julia and Elmer were the only people who called him Danny. It was a name for which great love poured out, as they considered him the eleventh grandchild. Dan and Elmer spent many hours together in a fishing boat, and he was working on "their farm."

"Rosie, we need to go to town right away," said Dan after hanging up the phone.

"For what?" Rosie asked.

"I'll explain on the way," said Dan.

Dan patched up the garage door as best he could, and Rosie had to have the toughest conversation to date with her parents. "Mom, you either need to hide the keys or you must give them to me."

Grandpa's driving days were over.

It was Mother's Day weekend.

Later that weekend, more challenging conversations took place when we discussed Grandma being able to care for Grandpa. This was a difficult ordeal for my mother as her father expressed extreme displeasure with his lot in life in some of his last lucid moments. We all cried.

Grandpa was moved to Rivers Bend Care Center in Manitowoc the following Friday. May, June and July did not go so well for anyone.

On July 6, Rosalie and Julia went to visit Elmer. "It was a wonderful day," recalled Rosalie. "He was himself one last time. Laughing and telling jokes," as he held a wood duck. "He even teased Mom asking, 'How did your hair get so gray?' We laughed some more."

One week later, Elmer passed on July 14, 2008.

While we were deeply sad for Grandma, she seemed to have more energy, no longer weighed down by Grandpa's earthly pain and her duties pledged on June 16, 1938, "Until death do us part."

EVERYONE CAME HOME

Four months later, November 1, 2008, was a joyous day at 525 Madison Street. Cars lined the short village block and every direct descendant of Julia Pritzl arrived at her home. Two days earlier, Julia had reached her ninetieth birthday. She flipped open Elmer's visitation book from his funeral a few months earlier, and on a page marked "Outdoor Memories" she wrote "Julia's 90th Birthday, Born October 30, 1918. Please sign my book." And so, some thirty-five family members, including children, grandchildren and great-grandchildren, signed her book.

Right before the family dined together, her son-in-law Randy asked Julia if she wanted to share what this day meant to her. Everyone went silent to hear Julia's quiet voice, as it no longer boomed out at full power. Very briefly she said, "Thank you God for Randy and Rosalie and for their care of *my* farm," with emphasis on the word *my*. Then she began to pray, "Bless us, Oh Lord, and these thy gifts which we are about to receive from thy bounty, through Christ, Our Lord. Amen."

Grandma had so much more she could have said that day. However, it solidified what I had long known since the first Ran-Rose cow, the Ran being Randy and the Rose being Rosalie, arrived on her family farm on May 7, 1981—Grandma Julia's heart had never left her farm.

With perfect attendance, four generations gathered to celebrate Julia (Burich) Pritzl's ninetieth birthday. The family matriarch is shown holding the youngest of her ten great-grandchildren on that day. *Author's collection.*

GRANDMA GOES HOME

Rosalie stopped by daily to assist Julia for nearly three years after her ninetieth birthday celebration. When Rosalie brought up the idea of looking at assisted living facilities in the summer of 2011, she was surprised that her mother acquiesced with little debate. Rosalie made appointments to tour some nearby accommodations.

The first trip took them to Manitowoc. Julia, with full mental capacities, surveyed the facility and deemed it too "fancy." That meant the always-thrifty Julia didn't like the price tag.

Rosalie made another appointment in nearby Brillion. When Rosalie went to pick her mother up for the 1:00 p.m. visit a week later, the door to Julia's home was locked.

"So, using my cellphone, I called Mom."

When she answered, Julia said, "Rosa, I had a heart attack." About the only time "Rosa" came out was when something was very wrong.

"Can you unlock the door?" replied Rosalie in an extremely concerned voice. "I'll try."

With that, Julia mustered her remaining strength and opened the front door to her home of the last thirty years. Then she fell into her daughter's arms. Julia would never open the door to her home again.

Rosalie called 911.

The ambulance arrived at 525 Madison Street. Halfway to the hospital, the ambulance pulled over on U.S. Highway 10. Rosalie instantly knew the paddles were coming out to revive her mother. The fast-acting paramedics had Julia's heart going again before Rosalie could get to the ambulance door as she forgot to let them know of her mother's wishes—DNR, do not resuscitate. However, Rosalie later learned that emergency medical technicians (EMTs) are required to revive patients in their care.

Julia arrived at the hospital and was seemingly resting comfortably, appearing asleep when a priest arrived, at Rosalie's request, to give Julia the Sacrament of Anointing of the Sick or what some refer to as Last Rites for Catholics. The young priest, just a few years out of the seminary, said, "Julia, you have lived a good life, you can go to Jesus."

To honor their mother's Bohemian baking traditions, Rosalie Geiger (née Pritzl) and Annie Krueger (née Pritzl) showcase kolaches destined for Julia (Burich) Pritzl's funeral dinner. *Author's collection.*

He paused, and that's when Julia, who hadn't spoken for hours, opened her eyes and said, "Please don't tell me when I should go see Jesus, I'm not ready yet. I'll go when I'm good and ready. But keep praying, I appreciate it." And with that, Julia went back to looking like she was in her final peaceful slumber.

Julia had one more moment of earthly feistiness that was her trademark and made her the family matriarch and the fourth generation to shepherd the family farm.

Clearly that shook both Rosalie and the priest. As the two went outside to talk, Rosalie pulled out a cigarette from her purse to have a smoke. The shaken priest asked for one, too, adding, "They don't prepare you for this in the seminary. Can I have one, too?"

Days later, on September 27, 2011, Julia departed her earthly life. Julia's family met one final time at her funeral on October 1, 2011, one month short of her ninety-third birthday. At the close of the gathering, her son-in-law Randy Geiger orchestrated the final funeral procession. Family and friends rolled past her beloved family farm as they ushered Julia to her final resting place in St. Mary's Cemetery next to her husband, Elmer, and her parents, Anna and John Burich.

They Saved Everything

That's when the heavy lifting ensued. Both products of the Great Depression, Elmer and Julia were ready for its second coming. We found bars of lead, stripped copper wires, waxed paper from cereal boxes and bags of twisty ties from bread containers all stored away. One day, the entire family, both children and grandchildren, came to help sort through things. Then they left. The task of stirring up memories was too much for most. A few returned from time to time. However, much work had to be done to clear the house for sale.

That's when my mother; my wife, Krista; and I became the weekly work crew that went about the task of cleaning out their home. That was a challenging task. As I had little expertise in making decisions on household items, I turned my attention toward the garage, where Grandpa and I had spent so much time over the past three decades.

Over two hundred hours into property cleaning, I had sorted through the first floor. It was now May 2012, and I needed to tackle the upstairs. "I need help!" I thought to myself on gazing at the next project. That's when I hired three neighborhood boys, two Amish and one "English" youth, with the term *English* being used in the Amish community when describing those not of their same religion. I also called Elmer John, Grandpa's only son. "Uncle Butch, can I have the lumber upstairs?"

"Sure, just ask your cousins if they want any. But they must haul it, not you. That should be yours for all the work you are doing." Cousin Chad took some maple boards to make some furniture. But the rest of the lumber came back to the home farm.

So, the job began with me and one of the boys up in the hot, humid attic. I was hot as if I were mowing hay in the peak of the barn. Sweat oozed out of every pore of my body. I picked up the first board. "Maple." I handed it to one of the boys working with me in the second story. "Maple," he repeated, carefully sliding the planed board out the window to the two young lads on the flat rack wagon. Their job was to separate the planks by each respective tree species.

"Maple."

"Hickory."

"Black cherry."

"Black walnut."

"Basswood."

"Ash."

"Red oak."

"White oak."

"White pine."

Single words kept emanating from my mouth in random order, as the pile had been disturbed over the decades as Grandpa dug for just the right piece of lumber. At times it sounded like I was a sergeant barking out commands to the privates.

Once the first wagon was filled, the boys hopped on the wagon and I drove the WD Allis-Chalmers, chugging back to the family farm to return the wood to where it began decades earlier.

We unloaded the lumber and then grabbed a bite to eat at the farm dinner table, served up courtesy of Mom. In some ways, it felt like a threshing crew had assembled once again as Mom made us a hearty meal fit for some hardworking farm hands. When our bellies were filled, the WD Allis-Chalmers engine came to life and heralded our journey back to the town of Reedsville for another load.

Upon backing the wagon carefully into the driveway, a small debate arose. Voices began to get louder, and I listened more intently. The boys were vying for a chance to go up in the second story.

These boards represented the final load of trees that came home to the family farm. The three young men who helped move those boards take one final rest before it was time to unload the last load of lumber. *Author's collection.*

"What?" I thought to myself.

"It's over 100 degrees in that attic. I don't even want to be up there, but I need to identify the timber," computed my brain.

"Why do you want to be up in that heat?" I asked out loud.

"To pick up lumber identification. Our dad and uncle will be impressed with what we are learning today," said the oldest Amish boy, who was intent on growing his knowledge of the trades.

I smiled and thought to myself, "These are Grandpa's kind of people—and mine, too."

I looked up and smiled to heaven, gave a nod to Grandpa, and with tears welling up in my eyes, I turned back to the boys and said, "You can take turns."

As the trees were coming home, Grandpa's carpentry shop came to life one last time. Class was in session. Only this time I was the teacher, and my students were awaiting their next instruction.

A BASIC FAMILY TREE
OF THOSE MENTIONED IN THE BOOK

Anna and John Burich (the third generation to run the farm)
<u>CHILDREN</u>
Mary Burich and her husband, Herb Kalies
Mary and Herb's son, Ken Kalies, is a central character in chapter 30
Agnes Burich and her husband, George Kubsch
Agnes and George's daughter, Janice Kubsch, is a character in chapter 5
Beatrice Burich and her husband, Quiren Sleger
Julia Burich and her husband, Elmer Pritzl
Cecilia Burich (died of polio in the eighth grade)
Wenceslaus Burich (died shortly after birth)
Adolphus Burich (died shortly after birth)

Anna and Louis Pritzl Sr. (parents of central character Elmer Wilfred Pritzl)
<u>CHILDREN</u>
Art Pritzl and his wife, Bernice (née Cummings)
Mary Pritzl and her husband, Walter Lashua
Veronica Pritzl and her husband, Earl Voss
Elmer Pritzl and his wife, Julia (née Burich)
Louis Pritzl Jr. and his wife, Lucille (née Woytasik)
Dorothy Pritzl and her husband, Larry Kubale

Julia and Elmer Pritzl (the fourth generation to run the farm)

<u>CHILDREN AND GRANDCHILDREN</u>

Jacqueline "Jacque" Pritzl, and her husband, Ken Konop, and children:
Julie and Chad

Elmer John "Butch" Pritzl and Martha (née Strauss) and children: Lisa,
Jody and Erik

Rosalie Pritzl and her husband, Randall "Randy" Geiger, and children:
Corey and Angela

Annette "Annie" Pritzl and her husband, Robert "Bob" Krueger, and
children: Samantha, Stephanie and Stacey

Rosalie Pritzl and her husband, Randall "Randy" Geiger (the fifth
generation to run the farm)

Corey Geiger and his wife, Krista Knigge (the sixth generation to run the farm)

BIBLIOGRAPHY

Aiello, Susan E. *The Merck Veterinary Manual*. Kenilworth, NJ: Merck, 2016.

Allis-Chalmers Corporation. *"60" All Crop Harvester*. Milwaukee, WI: Allis-Chalmers Manufacturing Company, 1948.

Apps, Jerold. *Rural Wisdom Time-Honored Values of the Midwest*. Amherst, WI: Amherst Press, 1998

———. *When the White Pine Was King: A History of Lumberjacks, Log Drives, and Sawdust Cities in Wisconsin*. Madison: Wisconsin Historical Society Press, 2020.

———. *Wisconsin Agriculture: A History*. Madison: Wisconsin Historical Society Press, 2015.

Austin, H. Russell. *The Wisconsin Story: The Building of a Vanguard State*. Milwaukee, WI: North American Press, 1948.

Baer, Marcie. "History Has a Familiar Ring." *Herald Time Reporter*, March 23, 1992.

Bergmann, Todd. "Nothing Hits Like Going Out for Fish." *Valders Journal*, March 10, 2022.

"The Big Dry." *Time*, July 1988.

Bradnan, Melinda. *Quality Czech Mushroom Recipes*. Cedar Rapids, IA: Museum Guild of the National Czech & Slovak Museum & Library, 1999.

Brillion High School press release. "Mrs. Konop Cited." *Post-Crescent*, December 1, 1972.

Busek, Martin. "Mushroom Gathering: Czech-American TV." Czech-American TV | Discover Czech Regions and Their Heritage, April 10, 2021. https://catvusa.com/traditions/mushroom-gathering/.

Cairns, C.A., Passenger Traffic Manager. *Chicago and Northwestern Line Time*, Table 36, page 30; Table 42, page 31, March 1928.

Calumet County Sheriff Department press release. "Area Snowmobile Accidents Kill Two." *Herald Times Reporter*, February 11, 1974.

Commercial West. March 14, 1908.

Cornell University. "Family Business Facts." https://www.johnson.cornell.edu.

Daniel, Kelly J. "White v. City of Watertown." Casetext, January 31, 2019. https://casetext.com.

Dose, Emmert. "Snowmobile Casualty Rate Tops Automobiles." *Journal Times*, February 11, 1974.

DuPont Corporation. *Farming with Dynamite: A Few Hints to Farmers*. Baltimore, MD: Lord Baltimore Press, 1910.

Dvorak, Norval. Obituary. *Herald Times Reporter*, October 8, 2015.

Fischer, Albert. Obituary. *Herald Times Reporter*, December 29, 1934.

Geiger, Corey. "I Want To Be Just Like Mom and Dad." *Hoard's Dairyman*, October 2017.

———. *On a Wisconsin Family Farm: Historic Tales of Character, Community and Culture*. Charleston, SC: The History Press, 2021.

Gurda, John. *The Making of Milwaukee*. Milwaukee, WI: Milwaukee County Historical Society, 1999.

Haas, Joanne. "DNR Warden: Italy's Hunt Shows Wisconsin's Safety Progress." Warden Wire, November 7, 2012. https://dnr.wi.gov.

Hammond, John. *Dutch Elm Disease Update, Agricultural Research*. United States Department of Agriculture, Agricultural Research Service, June 2006. https://agresearchmag.ars.usda.gov/2006/jun/elm.

Johnson, Janice. "Vest May Save Racers." *Post-Crescent*, December 8, 1974.

Kahler, Kathryn. "If It's Friday, It's Fish." *Wisconsin Natural Resources*, Spring 2022.

Kalies, Ken, Obituary. *Green Bay Press-Gazette*, October 10, 1974.

Krouse, Peter. "The Fall (from Dutch Elm Disease) and Rise of the American Elm: An Ohio Story." Cleveland.com, October 18, 2016. https://www.cleveland.com.

Kubale, Bernard. *The Place to Meet Your Friends*. Elabuk Press, 2017.

Ladd, Dave. *An Artist's Portfolio of Wooden Ducks*. Dodgeville, WI: Walnut Hollow Farm, 1983.

Lake to Lake press release. "Lake to Lake Merges." *Hoard's Dairyman*, January 10, 1981.

Lau, Chairman Diane. *History of Collins Wisconsin*. Collins, WI: Self-published, 1996.

Lee, John. "Snowmos Her Life…and Death." *Green Bay Press-Gazette*, February 12, 1974.

Lino, Mark, Kevin Kuczynski, Nestor Rodriguez and TusaRebecca Schap. *Expenditures on Children by Families, 2015*, United States Department of Agriculture, Center for Nutrition Policy and Promotion, Miscellaneous Report No. 1528-2015, January 2017. Revised March 2017. https://fns-prod.azureedge.us.

Lundsten, P.J. "White v. City of Watertown." Casetext, October 12, 2017. https://casetext.com.

Manley, Peter, Philip Freeburg, J.R. Habeck and Richard Stadelman. "Local Government Education." Local Government Education, December 2011. https://localgovernment.extension.wisc.edu/.

Moede, Albert, Obituary. *Herald Times Reporter*, November 14, 1940.

Moede, Richard, Obituary. *Herald Times Reporter*, February 26, 1985.

Moede, Ruth, Obituary. *Herald Times Reporter*, October 6, 1954.

O'Brien, Father John. *St. Mary's Parish Centennial Directory*. Brillion, WI: St. Mary's Parish, 1981.

Osman, Loren. *W.D. Hoard: A Man for His Time*. Fort Atkinson, WI: W.D. Hoard & Sons Company, 1985.

Pantzlaff, Andrew. "Brillion Man Ends Legendary Snowmobile Career." *Brillion New*, April 17, 2014.

Perry, Michael. *Population: 485.* New York: HarperCollins, 2002.

Potter, Julia. "Court Rules That Fence Law Applies to Cities and Villages." Boardman Clark Law Firm, December 12, 2017. https://www.boardmanclark.com.

Rebek, Eric, and Jennifer Olson. *Dutch Elm Disease and Its Control*. Oklahoma Cooperative Extension Service, EPP-7602. https://extension.okstate.edu/fact-sheets/dutch-elm-disease-and-its-control.html.

Rural telephone, *Special Assignment, Milwaukee Journal Stations Records*, December 29, 1967, Milwaukee Mss Collection 203, University of Wisconsin-Milwaukee Libraries, Archives Department.

Rusch, A.H. *Bee Keepers' Supplies*. Reedsville, WI: A.H. Rusch & Son Company, 1932.

Rusch, Arnold, F. Obituary. *Herald Times Reporter*, May 17, 1971.

Rusch, Arnold, H. Obituary. *Herald Times Reporter*, June 24, 1943.

Rusch, Reuben. Obituary. *Herald Times Reporter*, September 19, 1988.

Rusch, William. Obituary. *Herald Times Reporter*, August 20, 1966.

Scharenbroch, June. "Konop Inducted into Hall of Fame." *Brillion News*, March 5, 2009.

The Shawshank Redemption. Film. United States: Columbia, 1994.

Smith, Paul. "The Best Trophy of Wisconsin Hunting? Safety." *Milwaukee Journal Sentinel*, January 11, 2020.

Spielman, Patrick E. *Making Wood Decoys*. New York: Sterling, 1982.

Springer, William F. "Fences in Agricultural Wisconsin." Attorneys At Law-Brennan Steil S.C., October 30, 2017. https://www.brennansteil.com.

Staff, State Journal. "Outdoors: History of Deer Hunting in Wisconsin." Madison.com, November 15, 2011. https://madison.com.

Staff, Wisconsin Historical Society. "Logging: The Industry That Changed the State." Wisconsin Historical Society, August 3, 2012. https://www.wisconsinhistory.org.

Torgerson, Truman. *Building Markets and People Cooperatively: The Lake to Lake Story*. Arden Hills, MN: Land O'Lakes, 1990.

Tully, Grace. *Our Nation At War: Log of the President's Inspection Trip*. September 17 to October 1, 1942. Franklin D. Roosevelt Library and Museum.

United States Department of Agriculture, Farmer Cooperative Service. *News for Farmer Cooperatives.* April 1957, September 1959 and April 1962 editions.

United States Department of Agriculture, National Agricultural Statistics Service. "Wisconsin Dairy Cow Numbers, By Month, 1933 to 2022." October 2022. https://www.nass.usda.gov.

United States Department of Interior, Geographical Survey. *Chilton and Reedsville Quadrangle, Wisconsin, 15 Minute Series (Topographic)*. 1954.

Wendel, C.H., and George H. Dammann. *The Allis-Chalmers Story*. Sarasota, FL: Crestline Publishing, 1988.

Wenzel, Beth, Kris Bastian, Darcy Zander-Feinauer, Lisa Sprangers and Joey Diener. *Brillion Wisconsin, 1885–2010*. Brillion, WI: Zander Press, 2017.

Wisconsin Farm Bureau Federation. "County Farm Bureau Issue Backgrounder: Wisconsin Fence Law." September 2016. https://wfbf.com/.

Wisconsin Legislature: Chapter 90. https://docs.legis.wisconsin.gov/statutes/statutes/90.

World Series of Snowmobiling press release. "Brillion, Manitowoc Drivers Win." *Post-Crescent*, February 11, 1974.

———. "Brillion Snowmo Racers Win." *Herald Times Reporter*, January 21, 1974.

———. "Hunter Wins Two." *Green Bay Press-Gazette*, February 3, 1974.

———. "Three Area Snowmobilers in World Series." *Herald Times Reporter*, March 8, 1973.

Zachos, Ellen, and Douglas Merriam. *The Forager's Pantry: Cooking with Wild Edibles*. Layton, UT: Gibbs Smith, 2021.

Zander, Otto, and Elliot Zander. "Mrs. Louis Pritzl Meets Tragic Death." *Brillion News*, September 16, 1932.

Zarnoth, Dorothy, Editor. *History of Reedsville to 1976*. Brillion, WI: Zander Press, 1976.

ABOUT THE AUTHOR

C orey Geiger is the author of the award-winning book *On a Wisconsin Family Farm: Historic Tales of Character, Community and Culture.* Known to friends as the "Dancing Dairyman," Geiger danced his way onto Steve Harvey's *Family Feud* television show, Texas Two-Stepped on the colored shavings during the Supreme Champion ceremonies in front of a crowd of 4,500 at World Dairy Expo and competed at the Fred Astaire Cross Country Dance Competition national finals. There, he and his wife, Krista Knigge, won top amateur couple honors in their division.

In 1995, this University of Wisconsin–Madison graduate joined the *Hoard's Dairyman* editorial team and its publishing footprint dating back to 1870. In 2013, Geiger was tapped as the publication's fifth lead editor. Five years later, he spearheaded the effort that brought *Hoard's Dairyman China* to the marketplace, joining its English, Japanese and Spanish editions.

Geiger was elected the sixty-fifth president of Holstein Association USA and serves on the board of directors for the World Dairy Expo that drew attendees from eighty-six countries around the world in 2022. He was co-editor of World Dairy Expo's fiftieth anniversary book, *We Need a Show*, and co-chaired the fundraising campaign cabinet for the Farm Wisconsin Discovery Center, a $13.5 million educational facility. He has spoken in Canada, China, Mexico, the United Arab Emirates, Singapore and twenty-two U.S. states. He remains true to his farming roots, managing and owning a portion of his six-generation family farm, which serves as the epicenter for *The Wisconsin Farm They Built: Tales of Family and Fortitude*.

Other books by the author:
On a Wisconsin Family Farm:
Historic Tales of Character, Community and Culture

To schedule Corey Geiger
to speak about the book,
email: corey@coreygeiger.com